Investigating Physical Geography

Neville Grenyer

Oxford University Press

Oxford University Press, Walton Street, Oxford OX2 6DP

Oxford New York Toronto
Delhi Bombay Calcutta Madras Karachi
Petaling Jaya Singapore Hong Kong Tokyo
Nairobi Dar es Salaam Cape Town
Melbourne Auckland

and associated companies in
Beirut Berlin Ibadan Nicosia

Oxford is a trade mark of Oxford University Press

ISBN 0 19 913268 2

First published 1985
Second impression 1986

Filmset by MS Filmsetting Ltd, Frome, Somerset, UK.
Printed in Hong Kong

Acknowledgements

The publishers would like to thank the following for permission to reproduce photographs:

Aerofilms 2.9, 4.18, 6.14, 6.21, 7.1, 7:3, 7.8; Heather Angel 2.13, 5.16c, e, g, and h; Ardea 5.11, 8.7; A–Z Collection 5.16b; British Museum 5.7; Casella London Ltd 3.4, 3.15; J. Allan Cash 11.10; Cement and Concrete Association 2.19; T.J. Chandler 3.14; Bruce Coleman 5.16a, 9.12, 9.17, 9.18; Daily Telegraph Colour Library 4.1a, 11.15; Eastern Daily Press 10.11; Field Studies Council 7.12; Focal Point 8.14, 11.21; Ian Fraser 4.1e; Geoscience Features 1.14, 1.16a and b, 2.6, 5.12; Bob Gibbons 2.8, 9.7; Neville Grenyer 2.22, 6.24, 7.13, 9.1, 10.7; Andrew Goudie 9.16; The Guardian 10.15; Handford Photography 2.31; J.R. Hatfield-Powell 10.13; Houseman Ltd 2.12; Alan Hutchison 9.10; Illustrated London News 11.1; Institute of Geological Sciences 2.20, 6.13; Interfoto Archives 11.22; Eric Kay 2.4, 2.26, 4.5, 4.20, 7.7, 7.10; Frank Lane Agency 1.2, 5.16d and f; London Express News and Feature Services 11.18; Anthony Mann 6.16; Mansell Collection 1.1; Meteorological Office 4.11, 8.16; Munro 8.12; Octopus Books 3.9; Oxfam 9.11; Penrhyn Quarries Ltd 2.10; Picturepoint 3.6, p.88, 11.6; G.R. Roberts 6.1; John Shelton 1.20; Adrian Smith 6.8; Space Frontiers 3.11; Syndication International 6.12; Talentbank 9.19; Thames Water Authority 9.2; Jenny Thomas 2.1, 4.1d; Times Newspapers Ltd 10.2; United States National Hurricane Center 8.14; Victoria and Albert Museum 10.16; Vision International 1.19; West Air Photography 2.18, 10.4; Western Morning News 9.3; Jerry Wooldridge 2.16; J.W. Wright 4.1c; ZEFA 4.1b

Cover photograph by Adrian Smith

The following illustrations are by Roger Gorringe, Gary Hincks, Peter Joyce, Jon Riley, and Mike Saunders:

1.3, 1.13, 1.17, 1.22, 1.23, 1.26, 2.2, 2.5, 2.11, 2.15, 2.17, 2.25, 2.27, 2.28, 2.30, 3.1, 3.5, 3.8, 4.2, 4.4, 4.7, 4.8, 4.9, 4.10, 4.13, 4.15, 4.19, 4.21, 5.1, 5.2, 5.9, 5.10, 5.13, 6.2, 6.5, 6.7, 6.9, 6.10, 6.11, 6.15, 6.17, 6.18, 6.20, 6.25, 6.26, 7.2, 7.4, 7.5, 7.9, 8.1, 8.2, 8.3, 8.4, 8.5, 8.6, 8.11 (copyright *The Observer*), 9.15, 10.6, 11.3, 11.5, 11.16, 11.17, 11.19

Although the diagrams and maps have been drawn specially for this book, the author would like to acknowledge the following sources:
R. Ball, *The Earth's Beginning*, Cassell (1901), 1.25, 1.27; *British Regional Geology: London and the Thames Valley*, HMSO, 2.21; D. Brunsden and J.C. Doornkamp, *The Unquiet Landscape*, David & Charles, 1.13; A. Buck, *Teaching Geography*, 2, 3 (1977), 4.9; E. Bullard, *Transactions of the Royal Society*, 2.58A (1965), 1.9; T.J. Chandler, *Geographical Journal*, September 1962, 3.12 and *Weather*, 17 (1962), 3.13; R.J. Chorley, *Introduction to Physical Hydrology*, Methuen, 6.7; C.B. Cox and P.D. Moore, *Biogeography*, Blackwell Scientific, 5.6; C. Diver, *Geographical Journal*, 81 (1933), 5.15; I.G. Gass, *Understanding the Earth*, Artemis Press, 1.11; E. Gates, *Meteorology and Climatology for Sixth Forms*, Harrap, 4.3; Geography Department, Portsmouth Polytechnic, 10.5; G.H. Gimingham, *An Introduction to Heathland Ecology*, Oliver & Boyd, 9.5; C.L. Graves, *Mr Punch's History of Modern England*, Cassell, 11.9; K.J. Gregory and D.E. Walling, *Geography*, 56 (1971), 4.21; A. Hallam, *Scientific American*, November 1972, 1.22; K. Hilton, *Pattern and Process in Physical Geography*, UTP, 1.7, 1.15; A. Holmes, *Principles of Physical Geography*, Nelson, 1.5, 1.6; H.H. Lamb, *The British Climate*, English Universities Press, 8.15; C.M. Mason and M.E. Witherick, *Dimensions of Change in a Growth Area*, Gower Publishing, 10.9; F.E. Matthes, US Geological Survey, Professional Paper No. 160 (1930), 7.4, 7.5; R. Murray, *Meteorological Magazine*, 106 (1977), 9.4; *The Observer*, 1/2/70, 11.5; T.R. Oke, *Boundary Layer Climates*, Methuen, 2.30; V. Olgyay, *Design with Climate*, Princetown University Press, 3.5; P.W. Richards, *The Tropical Rainforest*, CUP, 5.17; R.J. Small, *The Study of Landforms*, CUP, 9.15; A.N. Strahler, *Physical Geography*, John Wiley, 4.8, 4.13; D. Weyman and V. Weyman, *Landscape Processes*, Allen & Unwin, 6.23; D. Weyman and C. Wilson, Geographical Association Occasional Paper No. 25, 4.15, 10.8; G.F. White, *Natural Hazards, Local, National and Global*, OUP, 10.14; J. Whittow, *Disasters*, Allen Lane, 11.20, 11.23; R.M. Wood, *On the Rocks*, BBC Publications, 1.18, 1.23, 2.5; A. Young and D.M. Young, *Slope Development*, Macmillan, 6.6.

Fig. 2.24 is reproduced by permission of the Director, Institute of Geological Sciences (NERC), NERC Copyright reserved/Crown Copyright reserved. The 1:50 000 Ordnance Survey map base is reproduced with the sanction of the Controller of HM Stationery Office, Crown Copyright reserved

Contents

To the teacher

Teachers have for some years been feeling uneasy about the physical geography taught in schools and the textbooks they have to use. The reasons for their discontent are many and varied, and some attempt is made to outline them here, together with a summary of the ways in which this book tries to overcome some of the problems.

Firstly, many physical geography textbooks seem to have failed to keep up with changes that have taken place in examinations. Most examinations demand more than rote learning and recall in physical geography; there is an increasing need to demonstrate geographical skills and reasoning with particular reference to decision-making. It is not sufficient for pupils to be passive assimilators in class; they need to undertake active investigation and to participate in enquiry.

Secondly, there has been concentration on physical geography as if it were isolated from humanity. Yet through developments in agriculture, industry, civil engineering and urbanization human activity has become a major influence on, for example, slope development, coastal processes, the composition of the atmosphere, and the world's climate.

Thirdly, the physical geography taught in schools bears very little relation to the environment in which pupils live and with which they should be concerned.

It has been argued that concentration on an exotic and different world provides its own justification, by presenting the landscape as an object of wonder away from the mundane and sordid problems of modern life. There is a need for wonder in our teaching, as there is a need for colour, to provide the basis of an appreciation of the beauty of the natural world and of the mighty forces that have shaped our continents and which move the atmosphere. It must be asked, however, whether the treatment of these in many textbooks provides that source of wonder or appreciation of beauty, or inspires any feeling of involvement or concern.

Modern teaching has a responsibility to explain scientific procedures, but it has a responsibility too to deal with attitudes and values. Many of our pupils are passionately concerned with their environment, yet in very few physical geography books is any guidance provided or any encouragement given to that concern.

In traditional geomorphology landforms have tended to be studied for their own sake. This has forced our teaching to be merely descriptive, for by ignoring the study of active processes there has been little opportunity for the pupil to seek explanations or achieve understanding. Neither has there been much chance to apply physical geography to the solution of problems or for the benefit of humanity and the quality of life.

It has been suggested that at school level the study of physical geography should be restricted to the inculcation of principles. However, such an approach would be meaningless without a prior appreciation of the reason for studying them and would deny pupils not intending to become professional geographers any insight into the applications of physical geography which provide much of the justification of the subject today.

Investigating Physical Geography aims:

1 To introduce pupils to the wonder of physical geography with colour photographs, maps and diagrams, and simulations where direct experience is impossible, e.g: the eruption of Krakatoa in Chapter 1; tropical cyclones in Chapter 8.

2 To relate physical geography to the urban and suburban environment with which most pupils are familiar, e.g: rocks and houses in Chapter 2; drainage on a housing estate in Chapter 10.

3 To provide opportunities for decision-making and involvement, e.g: taking decisions over issues in Exercise 16, Chapter 2; examining the working environment in Exercise 6, Chapter 3.

4 To demonstrate human influence on the environment and to inculcate a sense of responsibility, e.g: the mercury maze in Chapter 5; the destruction of heathland in Chapter 9.

5 To illustrate how ideas and theories have developed through the ages, e.g: the theory of plate tectonics in Chapter 1; plant succession in Chapter 5.

6 To help pupils develop skills of observation, mapwork, and recording and depicting data, e.g: the geological map in Chapter 2; the relief map in Exercise 2, Chapter 7.

7 To familiarize pupils with the vocabulary of physical geography; e.g: the summary exercises at the end of each chapter.

8 To encourage the study of geographical relationships, e.g: by correlation as in Exercise 7, Chapter 10; by using flow diagrams as in Exercise 11, Chapter 9.

9 To illustrate the processes at work shaping the environment, e.g: river processes in Chapter 4; pollution in Exercise 3, Chapter 11.

10 To encourage simple, practical fieldwork and experiment in the local environment using easily constructed equipment, e.g. home-made rain gauges in Chapter 4; evaporation studies in Exercise 11, Chapter 4.

11 To introduce pupils to model-building and simulation, e.g: the model of a depression in Exercise 9, Chapter 4; the spit growth simulation in Chapter 6.

12 To provide opportunity for simple enjoyment and fun in physical geography, e.g. the mercury maze game in Exercise 4, Chapter 5; the round the world yacht race in Chapter 8.

1 Earthquakes and volcanoes

Experiencing an earthquake

Here are two extracts from books describing what it was like to be caught in an earthquake. The first was in Lisbon in Portugal in 1755.

'There were three major shocks that day in Lisbon and hundreds of minor shocks as the lurching earth settled to take up new pressures.

'The second shock lasted about three minutes. From the River Tagus, where sailors on the anchored ships had watched with horror the first collapse, a fantastic sight was presented. The master of one English ship saw the whole city waving backwards and forwards like the sea when the wind first begins to rise.

'Lisbon is built on many sharp hills which at the time projected a great number of convent and church spires into the sky. The master of another ship compared their movements to the waving of a field of corn in the wind.

'Then, out in the estuary, he saw a high bank of water rising as if the sea were forming into a mountain. Foaming and roaring, the huge wave swept towards the shore at great speed. As the crowd on the waterside understood what was happening they broke and ran, anywhere away from

Fig. 1.1 The great Lisbon earthquake: all the events of that day are shown as if happening at once. You can see the three types of danger referred to in Exercise 1 on page 6

the water, back into the devastated city. But the hissing waters caught them.

'After the third shock the great fire had started. It was to burn out all the centre of Lisbon. It was to continue for nine days.'

(Allen Andrews: *Earthquake*, Angus & Robertson)

1 Name three types of danger described in the extract.

The second earthquake took place in Chile in the 1930s.

'First a rapidly approaching rumble like the sound of a half-loaded lorry bounding along a rough road; then a slight shake followed by a pause. Another bigger shake would follow, during which eyes were riveted on the movement of water in a bowl of flowers, or pictures on a wall.

'Pause and shake continued, reaching a climax and then subsiding at the same tempo, during which everyone sat in a sprinting position ready to make for the safety of open country.

'Once and once only did I actually see an earthquake, when Marie and I were riding through a dried-off field, bare of grass. I had dismounted to fix a spur and as I knelt to adjust it, I heard that ominous rumble.

'Looking towards the sound, at that low level, I saw a wave coming towards me across the field, just as one travels over the surface of the sea. About one foot in height, it advanced and passed me at great speed, throwing me off balance, while both horses stood with legs splayed out like milking stools to keep their feet.'

(C. J. Lambert: *Sweet Waters*, Chatto & Windus)

Fig. 1.2 Earthquake damage in Alaska: the angles of the houses and trees show what has happened

2 What are the main signs that an earthquake is happening?

3 What do you suppose are the main dangers of being caught in an earthquake:
a) in a big city?
b) in mountains?
c) by the coast?

Measuring earthquakes

The strength of an earthquake is measured by the amount of energy it gives out to shake the land. This is what is measured by the **Richter scale** which is usually quoted by newspapers and television to give an idea of the size of the earthquake.

The Richter scale runs from 0 to 9. It is easy to show what an earthquake of scale 1 feels like if someone jumps onto the floor from a metre-high table. A scale 2 earthquake releases ten times as much energy as a scale 1 quake. Scale 3 would have a hundred times the energy of scale 1.

4 How many more times more powerful than scale 1 would an earthquake be of:
a) scale 4?
b) scale 5?
c) scale 8?

The effect of an earthquake depends on where it happens. The earthquake in Chile was probably as strong as the Lisbon earthquake (pp. 5 and 6), yet because it was in the open countryside it did little or no damage.

The effect of an earthquake can also be measured on the Modified Mercalli scale (Fig. 1.3).

5 Explain in your own words the differences between the two ways of measuring earthquakes.

1. Detected only by instruments.

2. Very feeble – noticed only by sensitive people.

3. Slight – felt by people at rest.

4. Moderate – felt by people who are moving about.

5. Rather strong – people wake up, bells ring themselves.

6. Strong – slight damage to buildings.

7. Very strong – walls crack, people panic.

8. Destructive – chimneys fall.

9. Ruinous – houses fall down.

10. Disastrous – many buildings destroyed.

11. Very disastrous – few buildings left standing, ground cracks.

12. Catastrophic – total destruction of buildings, ground badly twisted.

Fig. 1.3 The Modified Mercalli scale of earthquake intensities runs from 1 to 12

The black dots are minutes

P S L

Fig. 1.4 A seismograph record of an earthquake showing P, S, and L waves

The world pattern of earthquakes

Shock waves from earthquakes are recorded on an instrument called a **seismograph**. If we draw a line through all places that feel an earthquake with the same strength, it will form a circle or an oval. The centre of this ring will be the place where the earthquake started: the **focus** or **epicentre** of the earthquake. Fig 1.4 shows a record made by a seismograph.

Three main sets of waves are shown on a seismograph record of an earthquake. The first or preliminary shock waves (P waves) are followed by a second set of waves (S waves). Both P and S waves have passed directly through the earth and are followed by a longer set of waves (L waves) which have come the long way round the surface of the earth.

Because P waves travel faster than S waves, the gap between them gets larger the further away they are from the earthquake focus as shown on Fig. 1.5. We can use Fig. 1.5 to tell us how far away the focus is if we know the length of time between the arrival of the P waves and S waves (Fig. 1.4).

☆**6** How far away was the focus of the earthquake from the seismograph? Work this out using the gap between the P and S waves in Figs. 1.4 and 1.5.

As long as we have three seismograph records we can find out exactly where the focus of the earthquake lies, as shown in Fig. 1.6.

7 a) Fig. 1.8 is a list of all the *major* earthquakes that have occurred between 1934 and 1964. Each member of the class should choose an earthquake and then use an atlas to locate the place where the earthquake occurred.
b) Mark these places on a large class wall map or an overhead projector outline map of the world.
c) Describe the pattern shown on the map.

8 Fig. 1.7 shows *all* the earthquakes that occurred between 1961 and 1967.
a) Compare it with the map produced for Exercise 7.
b) Do some areas which have earthquakes marked on one map, not have them marked on the other?

☆ Exercises marked with a star are more difficult.

Fig. 1.5 The time taken by P, S, and L waves to travel up to 11 000 km

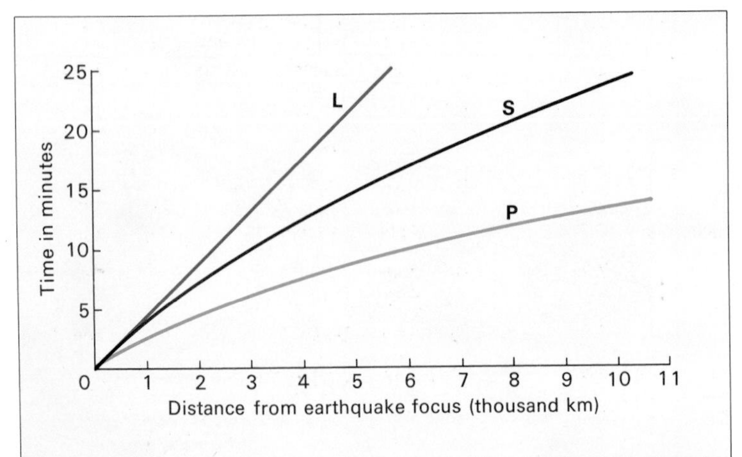

Fig. 1.6 Plotting an earthquake's focus using three seismograph records

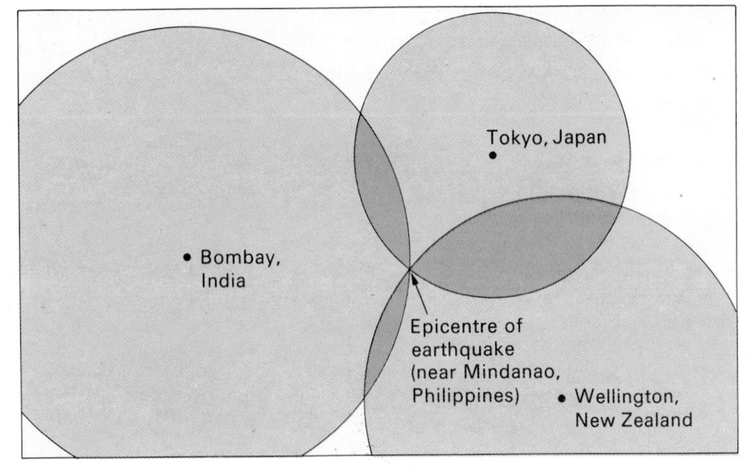

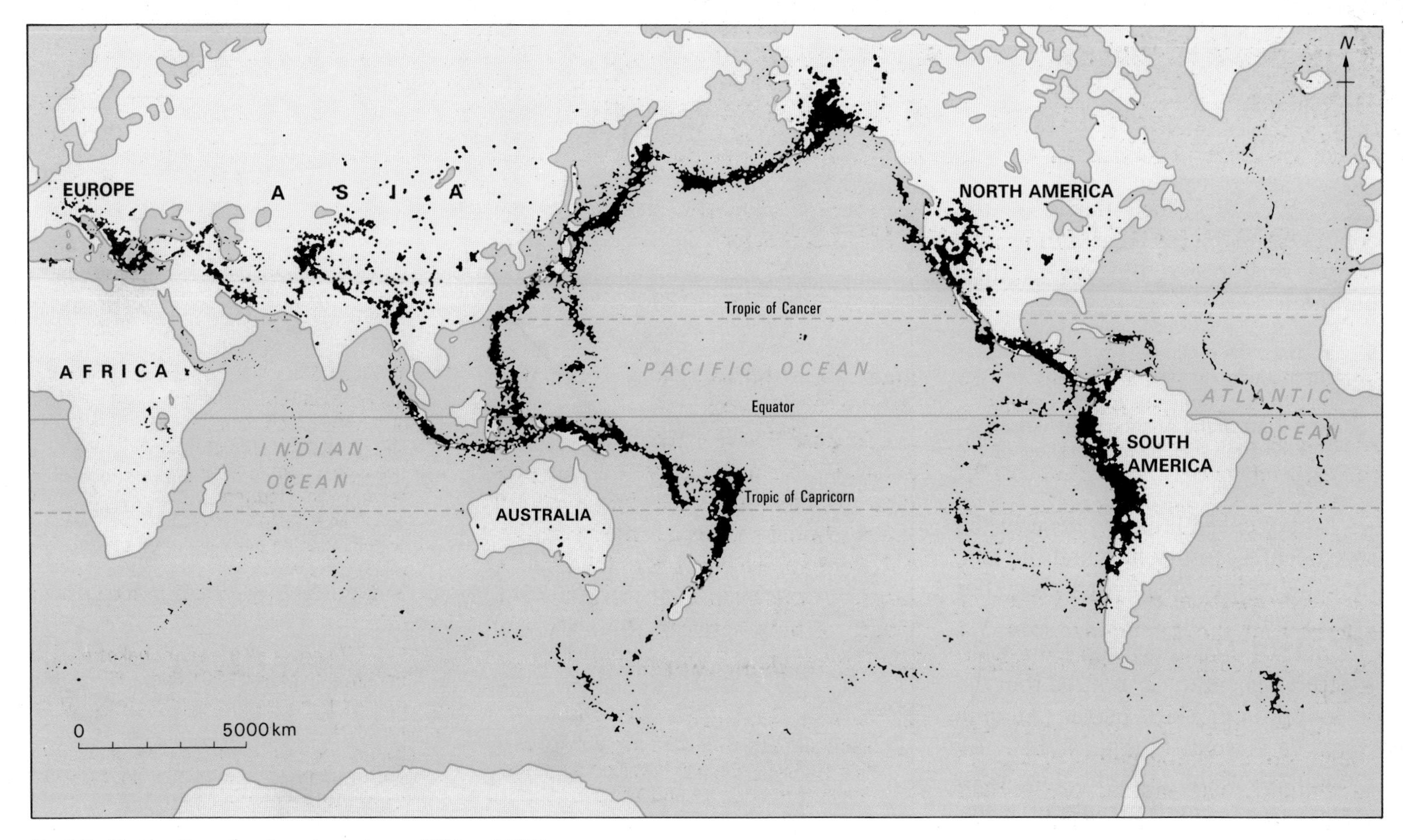

Fig. 1.7 World pattern of earthquakes between 1961 and 1967

Fig. 1.8 Major earthquakes over a period of 30 years

Date	Epicentre	Richter scale
1934 Jan	Bihar, India	8·4
July	Santa Cruz Is.	8·1
1935 May	Quetta, Pakistan	7·5
Dec	Sumatra	8·1
1938 Feb	Java	8·6
1939 Dec	Anatolia	7·9
1941 June	Burma	8·7
June	Central Australia	6·8
Nov	West Portugal	8·4
1942 May	Ecuador	8·3
Aug	Guatemala	8·3
Aug	Brazil	8·6
Nov	To the south of Africa	8·3
1943 April	Andes	8·3
1944 Dec	Honshu, Japan	8·3

Date	Epicentre	Richter scale
1946 Aug	West Indies	8·1
Dec	Shikoku, Japan	8·4
1949 Aug	South Alaska	8·1
1950 Feb	Hokkaido, Japan	7·9
Aug	Assam	8·7
Dec	New Hebrides	8·1
Dec	Andes, Argentina	8·3
1952 July	Kern County, California	7·7
1953 Mar	NW Anatolia	7·2
1954 Sept	El Asnam; Algeria (Orleansville)	6·8
1955 Feb	Quetta, Pakistan	6·8
1956 Dec	Baja, California	6·8

Date	Epicentre	Richter scale
1957 Mar	Andreanof Is	8·0
April	To the south of Samoa	8·0
July	Guerrero, Mexico	7·5
1958 July	SE Alaska	8·0
1959 Jan	Brittany	5·2
1960 Feb	North Algeria	5·5
Feb	Agadir, Morocco	5·8
April	Lar, Iran	5·8
1961 June	Ethiopia	6·8
Aug	Peru/Brazil	7·5
1962 Sept	Buyin, Iran	7·5
1963 July	Skopje, Yugoslavia	6·0
1964 Mar	Anchorage, Alaska	7·5

The continental jigsaw

Fig. 1.9 shows the shapes of the continents using the edge of the **continental shelf** below the sea as their outline.

9 a) Trace the outlines from Fig. 1.9 and cut them out.
b) Join the continents together so that both sides of the Atlantic meet almost exactly. Then stick them on a sheet of paper.

A man called Francis Bacon in the seventeenth century first had the idea that the continents might fit together like a jigsaw.

In 1801 Alexander von Humboldt, a German geographer, noticed that the rocks on both sides of the Atlantic were very similar (Fig. 1.11). He thought that the Atlantic had been formed by surging currents of water which washed away the land between the continents.

In 1885 Edward Suess, an Austrian geologist, put forward his idea that the continents might at one time have been joined together as a huge mass of land called Gondwanaland. He thought that the areas between the present continents might have sunk to form the ocean floors (Fig. 1.12).

The German, Alfred Wegener was the first person to suggest that the continents might have **drifted** apart from a central land mass. When his ideas were published in 1924, geologists and physicists objected most strongly. They picked on Wegener's weakest theory: his idea about what caused the continents to move. Wegener thought that the spinning of

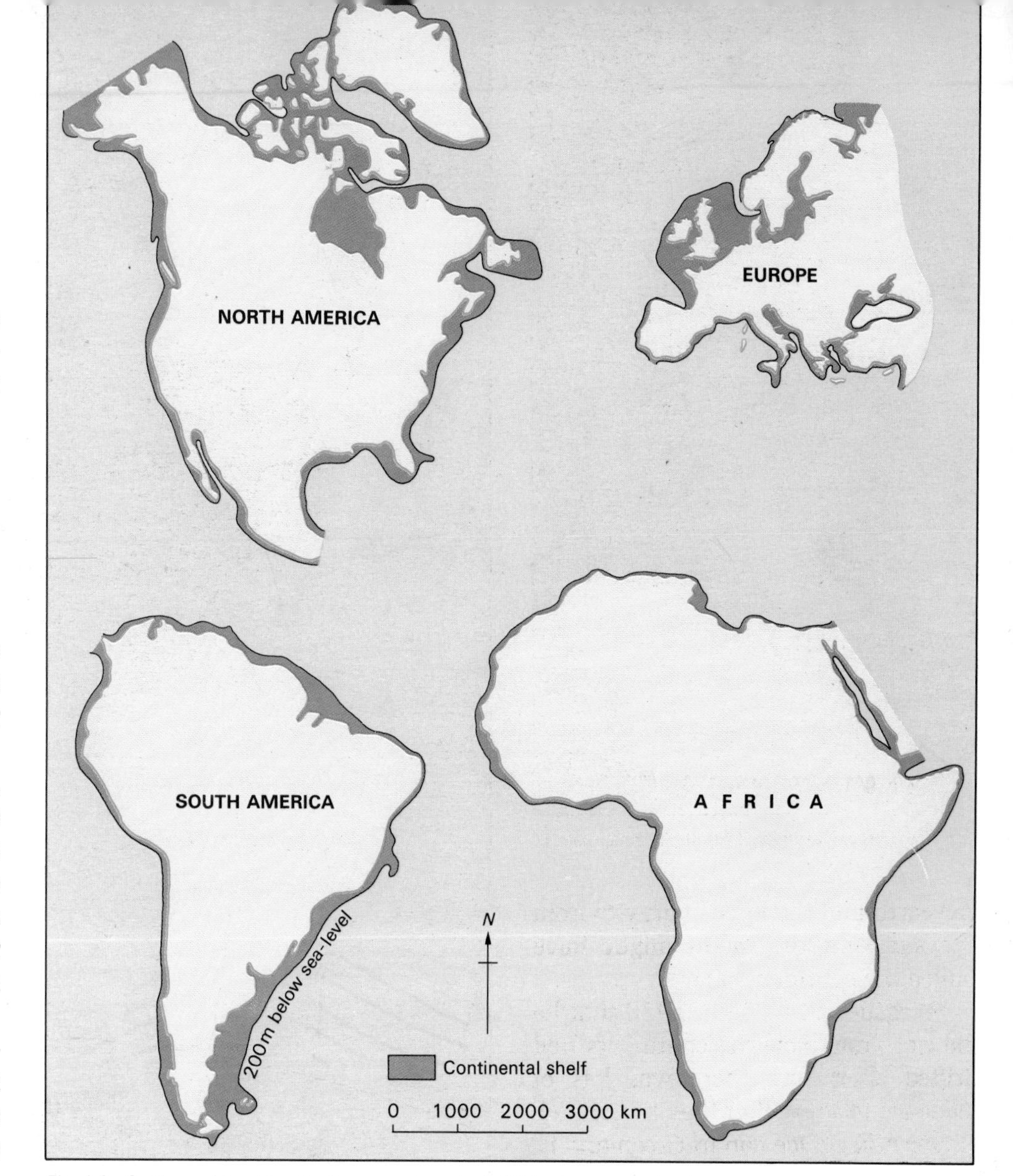

Fig. 1.9 Outlines of the Atlantic Ocean continents at 200 m below sea-level

Fig. 1.10 Table to sum up the growth of ideas about continental drift

Date	Scientist's name	Idea	Explanation
1620	Francis Bacon	Matching coastlines of continents	None
1801	Alexander von Humboldt	Similar rocks and geology in Africa and South America	Land between Africa and South America had been washed away
1885	____________	____________	____________
1924	____________	____________	____________
1944	Arthur Holmes	Convection currents	____________

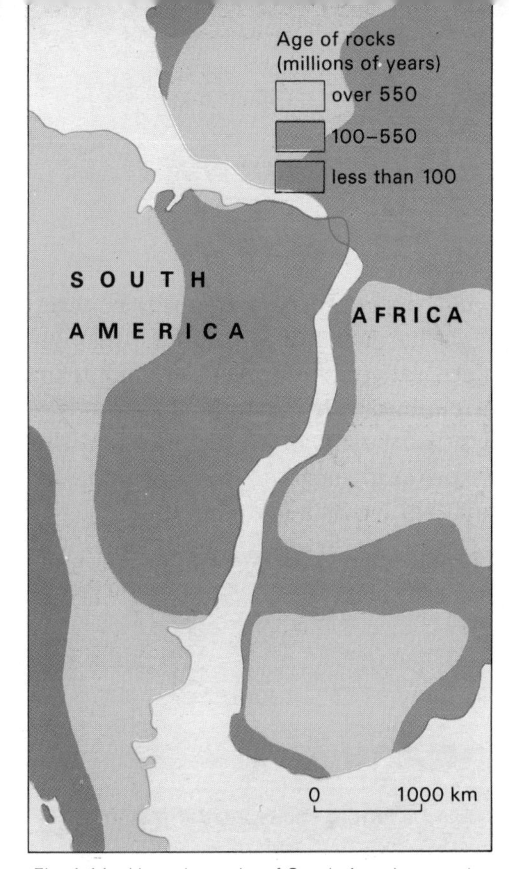

Fig. 1.11 How the rocks of South America match those of Africa

the earth and the force of gravity from the sun and the moon might have pulled the continents apart.

Wegener admitted in 1928 that he did not know *how* the continents had drifted apart, but there was lots of evidence to show that they *had* drifted. He died on an ice cap in Greenland in 1930 still looking for evidence to support his ideas.

Arthur Holmes, a Scottish geologist, kept Wegener's idea alive. Holmes argued that convection currents in the molten rock inside the earth could cause the crust to move about and so to move the continents. Very few other geologists agreed, however, and by 1950 the idea of continents drifting was treated like a geological joke.

10 Sum up the growth of ideas about continental drift by finishing the table in Fig. 1.10.

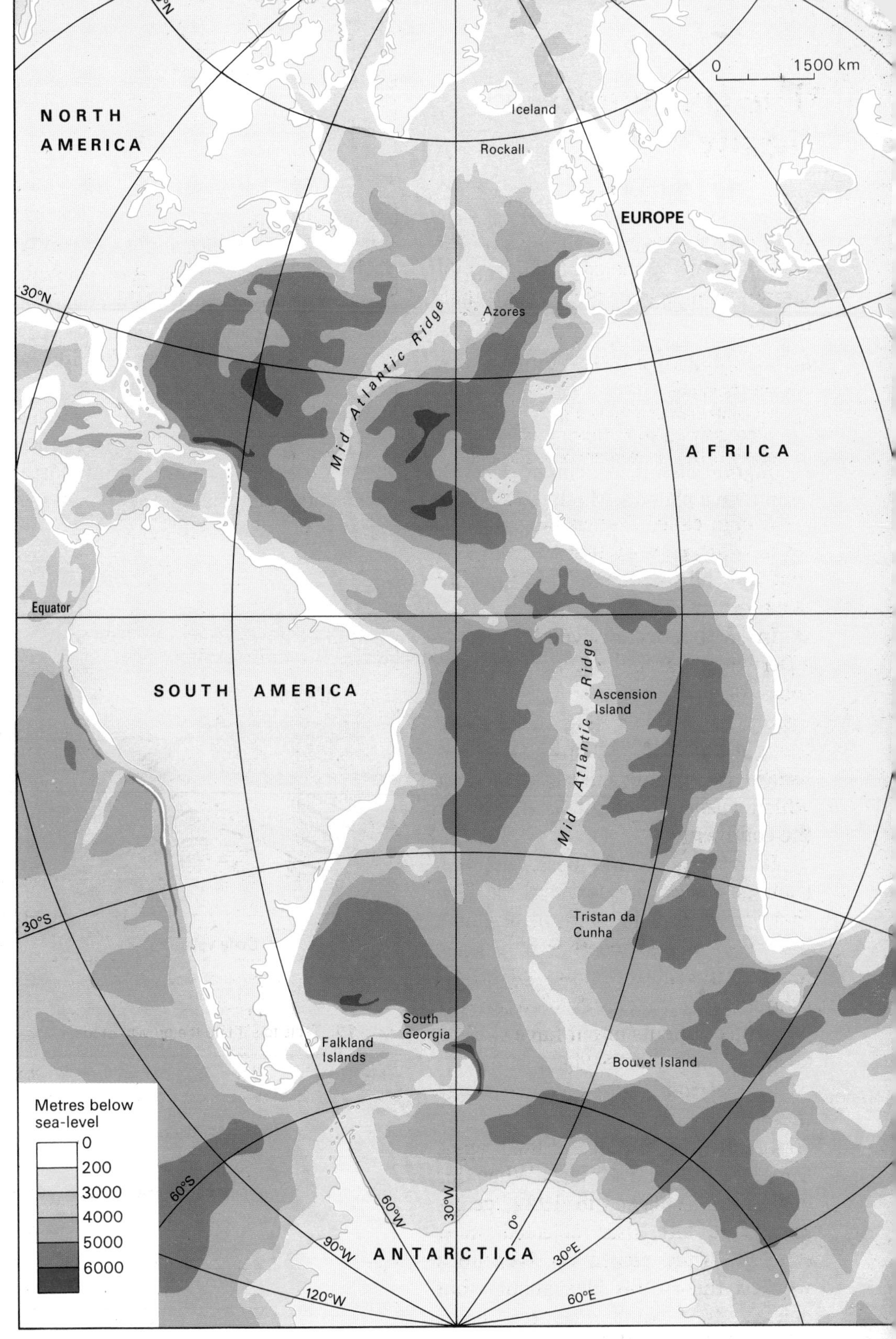

Fig. 1.12 The floor of the Atlantic Ocean

Types of volcanoes

Molten rock is only known as **lava** when it has come out of the ground. Molten rock still in the ground is known as **magma**. The type and shape of a volcano depends mostly on the type of magma and the shape of the hole out of which the magma comes (Fig. 1.13).

If very fluid basaltic magma erupts through a fissure, it spreads out to form a **lava plateau**. Much of Northern Ireland is an example of a lava plateau. If basaltic magma erupts through a tube vent, it forms a low **shield volcano** which looks like a viking soldier's shield lying on its back.

Intermediate magma erupting through a central tube produces a **cone-shaped volcano**. If the central vent becomes blocked, smaller vents often break out around it and form small cones on the side of the main one. This is called a **composite cone**.

When acid magma erupts through a central vent, it forms a **cone** made up of **clinker**. When it erupts through a fissure, the magma quickly plugs the fissure and forms a swelling dome of lava known as a **lava dome**.

11 Using Fig. 1.13, list the types of magma and the openings through which they reach the surface.

Fig. 1.13 The different types of volcanoes

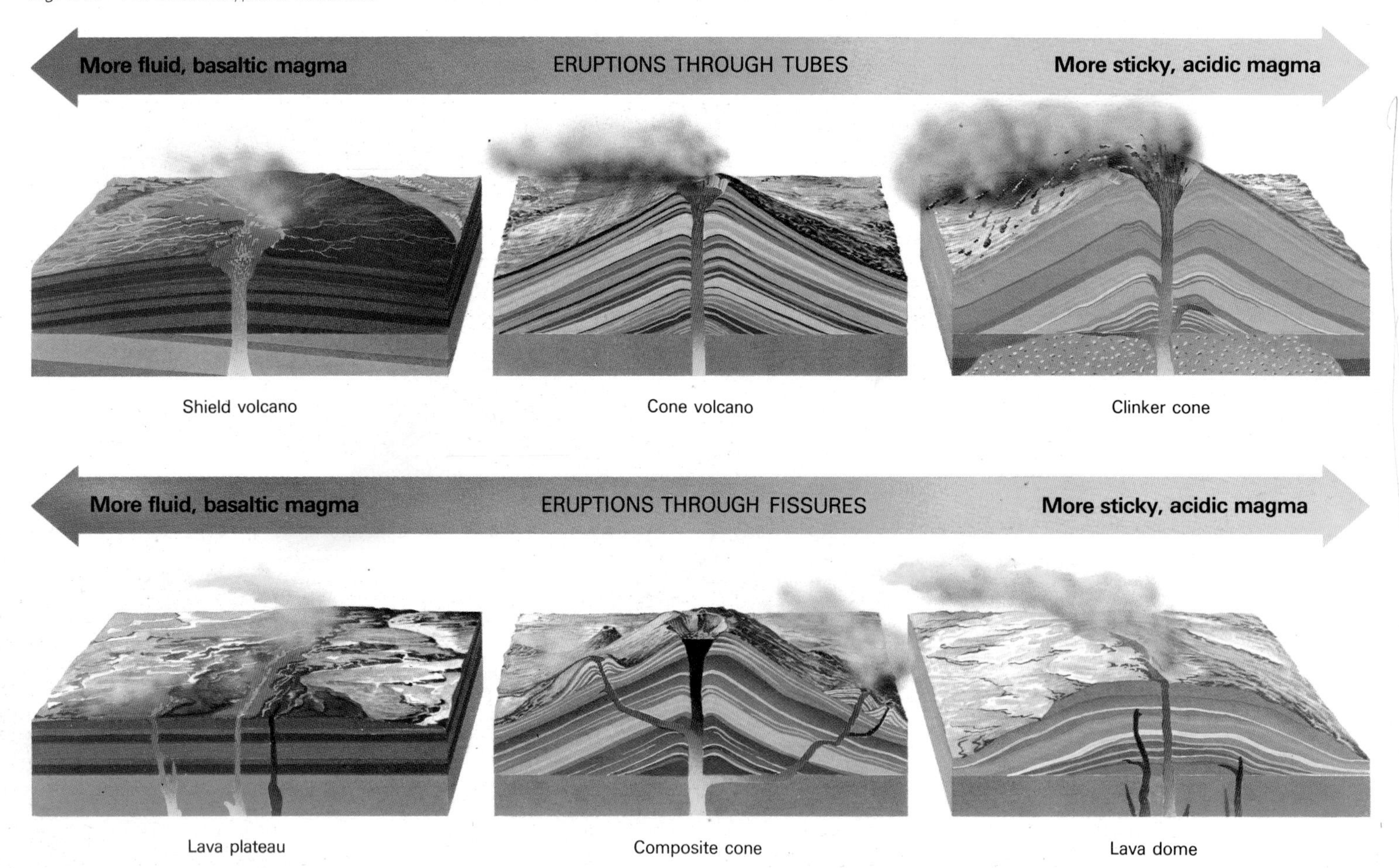

Volcanoes in Iceland

The geologists of Iceland are very lucky because in their small island there are examples of most of the different types of volcano.

12 What types of volcano are not shown in Fig. 1.15?

On 14 November 1963 the cook on a fishing-boat south of Iceland noticed black clouds rising from the surface of the sea. Within twenty-four hours a new island had been formed, to be called the island of Surtsey (Fig. 1.14).

Fig. 1.14 The island of Surtsey, 70 metres high a week after it first appeared

Fig. 1.15 Iceland: geology and volcanoes

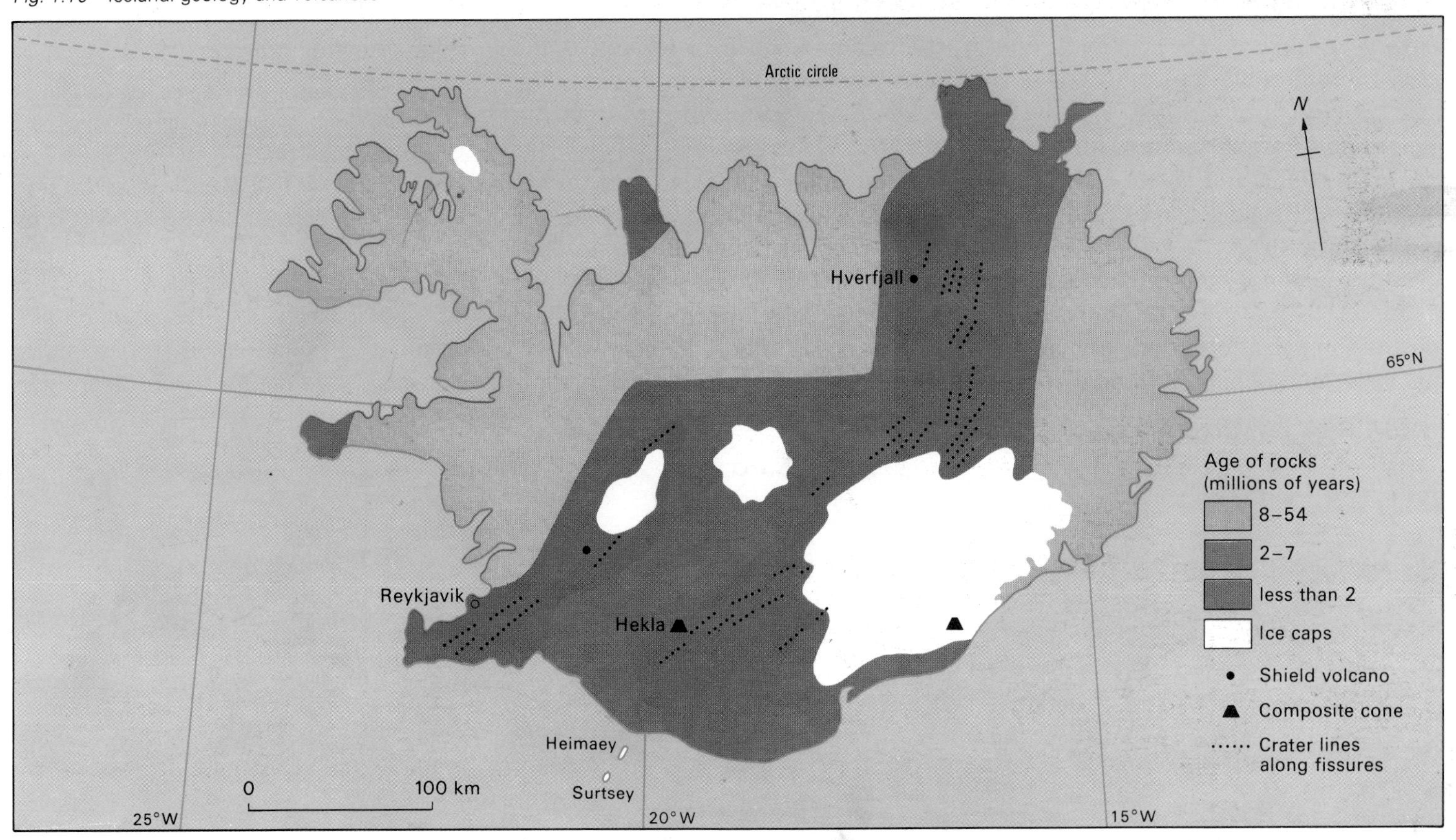

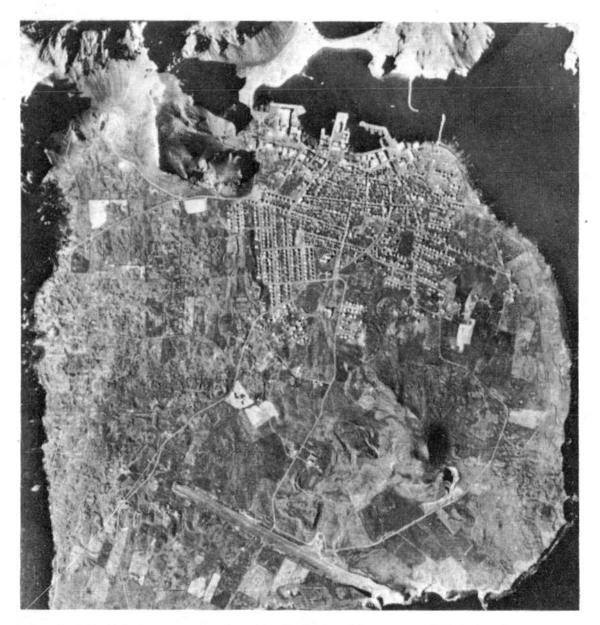

Fig. 1.16 Heimaey, Iceland: *(left)* before and *(right)* after the eruption. Find the jetty in both photographs to see how much new land was created

On 23 January 1973 on the island of Heimaey, only 15 km north of Surtsey, a fissure $1\frac{1}{2}$ km long appeared. From along its length red hot lava shot up into the air. Fig. 1.16 shows how much more land was created on the island north-east of the old volcano Helgafell which had been quiet for 5000 years.

Surtsey and Heimaey are shown on Fig. 1.15. This map also shows that most of the recent eruptions are creating new land where Iceland's rocks are youngest, while the older rocks are found on either side.

Stretching south of Iceland is the underwater Mid-Atlantic ridge (Fig. 1.12). In the 1960s Harry Hess, an American geologist, put foward his idea that **mid-ocean ridges** are on top of rising convection currents in the molten rock far below the ocean floor (Fig. 1.17). Out of the centre of these ridges new rocks flow from volcanoes.

The ocean floor is very slowly spreading sideways in both directions from the central ridge. Iceland, perched on top of the Mid-Atlantic ridge, is being pulled apart.

13 Add a note about Hess's ideas to your table from Exercise 10 summarizing the history of the idea of continental drift.

☆**14** Look at Fig 1.12 carefully. Are there any other islands which you might expect to behave in the same way as Iceland?

According to Hess, where the ocean floor reaches the edge of the ocean it is pulled down into what is called a subduction zone near the

Fig. 1.17 Hess's idea of ocean floor spreading

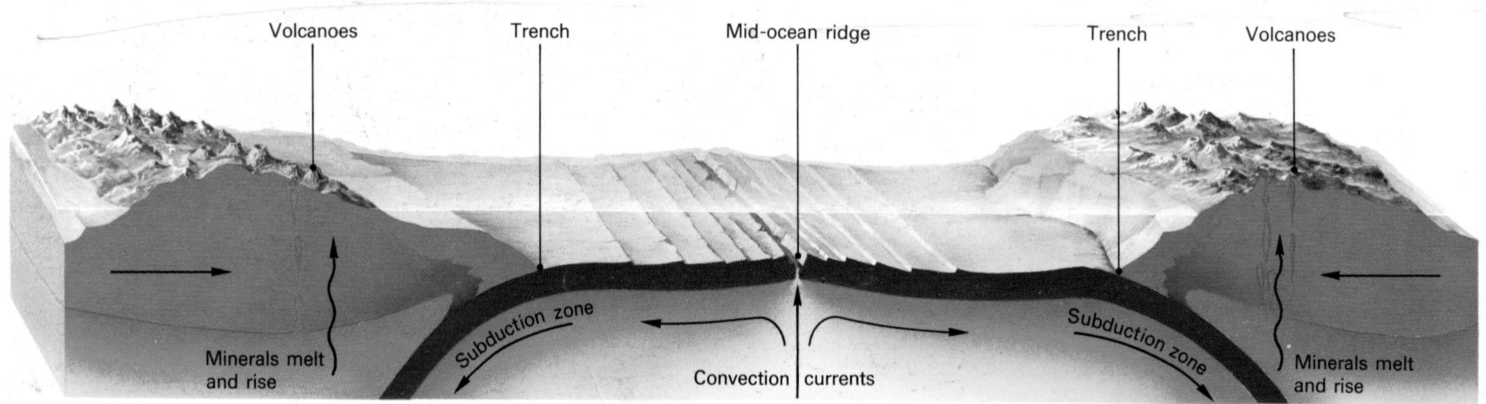

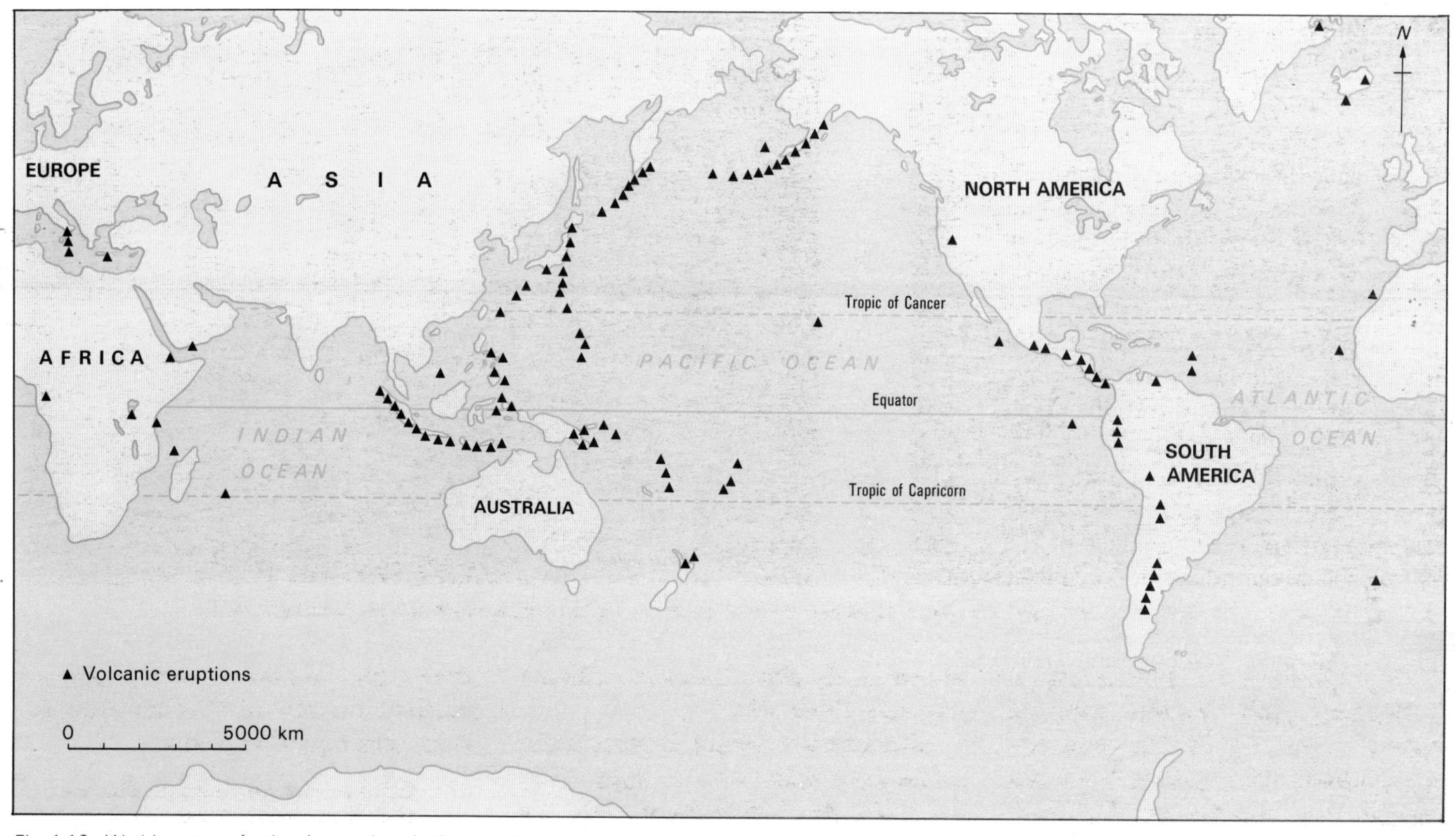

Fig. 1.18 World pattern of volcanic eruptions in the twentieth century

ocean trenches (Fig. 1.17). Where the crust gets dragged down, some of the minerals melt and expand rapidly, rising to the surface as more volcanoes.

15 Fig. 1.18 shows all the volcanoes that have erupted in the world this century. Use your atlas to give names to as many as you can. Mark their names on a copy of Fig. 1.18. Use newspaper and television reports to keep your record up to date.

16 Use chalk or pastel crayon to mark the pattern of earthquake and volcanic belts onto a globe. What are the main differences between the patterns shown on your globe and those shown on maps like Figs. 1.7 and 1.18?

Fig. 1.19 A house on Heimaey buried in volcanic ash from the 1973 eruption

Plate tectonics

Fig. 1.20 River courses deflected by the San Andreas fault (running from left to right)

In 1968 W. J. Morgan put forward his idea of how the continents moved and people began to realize how close to the truth Holmes's and Wegener's ideas had been. Morgan suggested that the earth's crust is divided into a number of rigid plates which move slowly in different directions (Fig. 1.21). This idea is known as the theory of **plate tectonics.**

The regions of earthquakes that we saw in Fig. 1.7 are the edges of the plates or **plate boundaries.** Some plates

Fig. 1.21 World pattern of plates and plate boundaries

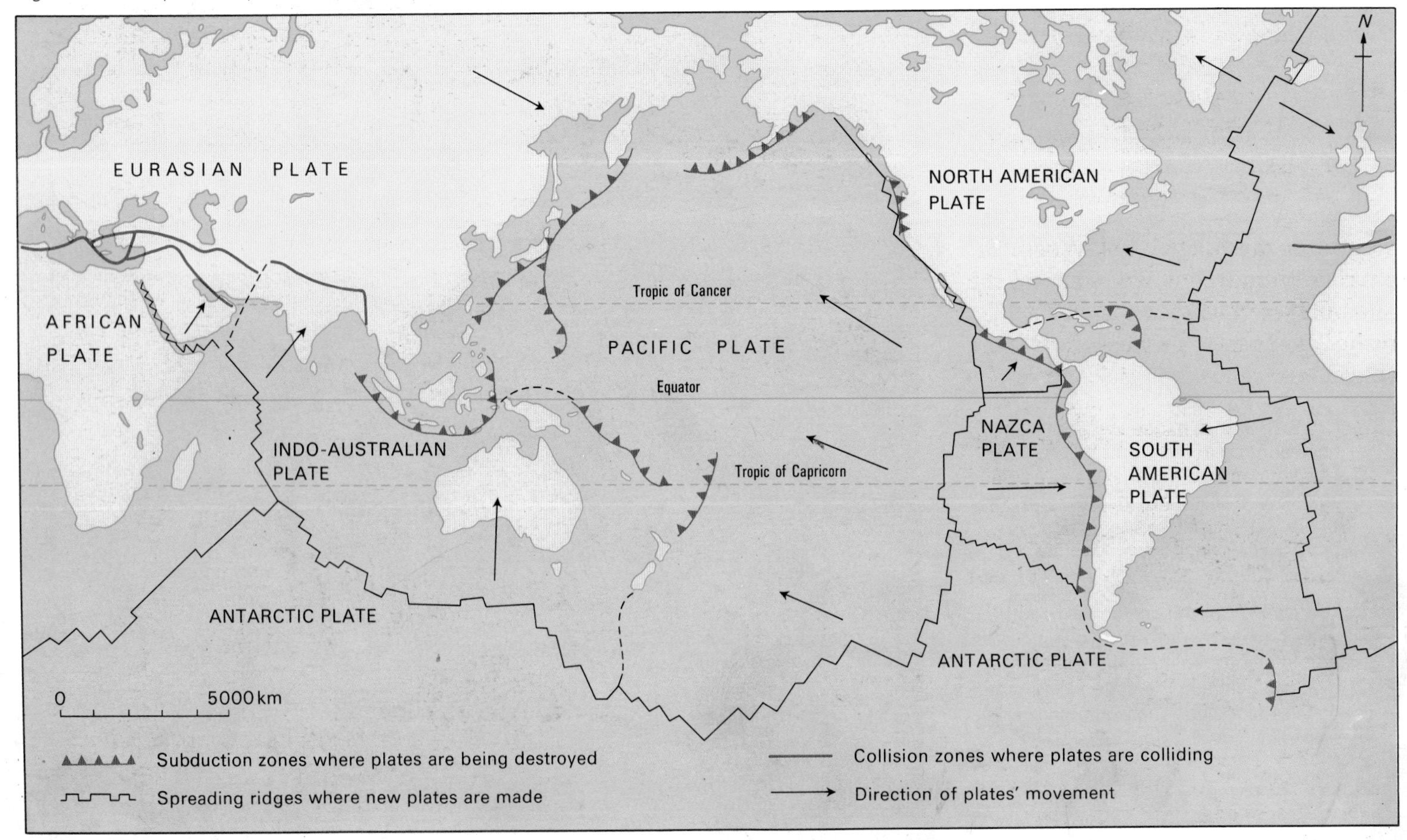

are moving sideways past each other like the San Andreas fault in California (Fig. 1.20). The mid-ocean ridges (Fig. 1.17) are plate boundaries where new plates are being formed, while the subduction zones are where plates are being destroyed.

In some areas such as the Mediterranean two or more plates are colliding and crumpling at the edges to form fold mountains; these are called **collision zones**. Most of the world's earthquakes are the results of stress at plate boundaries. Some earthquakes are caused by a new break or **fault** being made in the rocks of the crust. Most take place along old fault zones.

17 Look again at the map you produced of earthquakes in Exercise 7. What kind of plate boundaries are found at each of the earthquakes you plotted? Use Fig. 1.21 to help you.

☆**18** Describe in your own words how the idea of plate tectonics explains:
a) why some continents can be fitted together.
b) where earthquakes occur.

Fig. 1.22 tries to show what happened to the animal life when North and South America were joined together 22 million years ago. Before the continents were joined there were two separate groups of animals. Afterwards, one of the South American animals migrated to the north while a number of North American animals went the other way. Some found the competition of new animals too great and became extinct.

19 Look at each animal in Fig. 1.22 and decide whether it moved, stayed where it was, or became extinct.

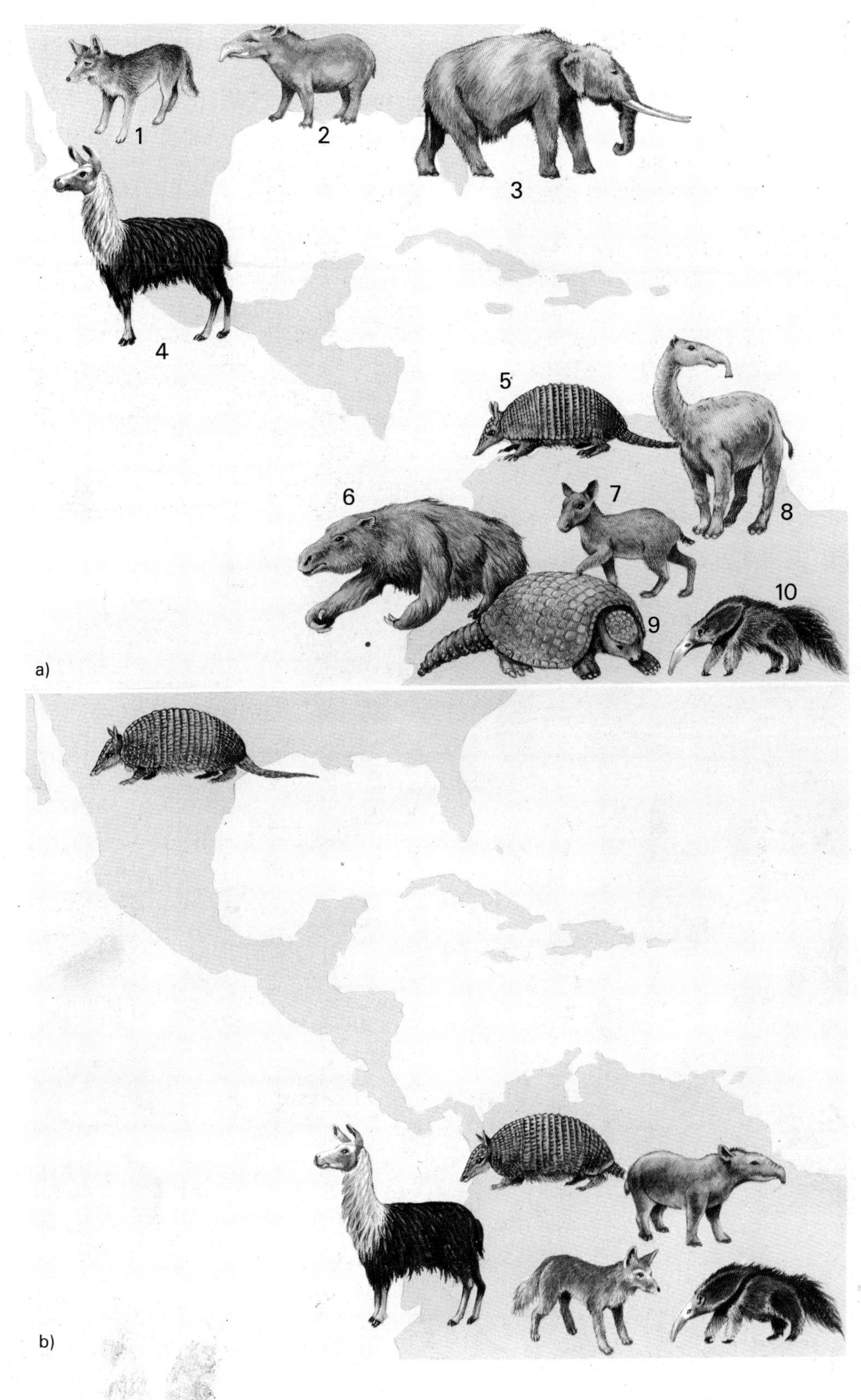

Fig. 1.22 Animals found in North and South America: a) before they were joined, b) afterwards

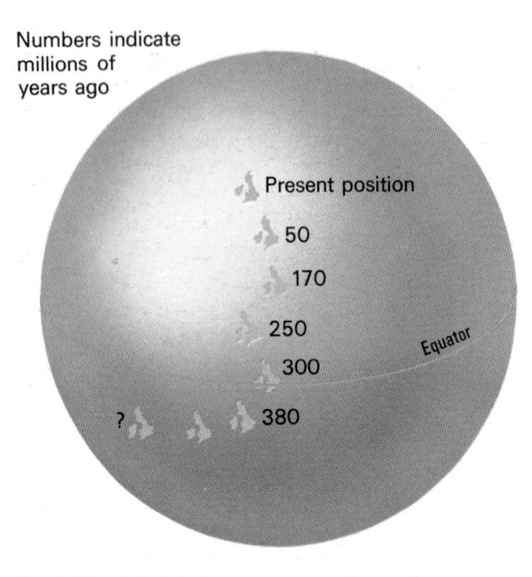

Fig. 1.23 Britain's journey across the earth

Because the various plates have moved, Britain has been carried about on the globe as shown in Fig. 1.23. Many rocks contain minute grains of a mineral called magnetite that shows where north stood at the time they were formed. It seems that the earth's climatic zones have remained in roughly the same place and that a rock such as limestone, made from tropical corals, was actually laid down in the tropics and has been moved to its present position.

Fig. 1.24 shows how long ago the various rocks of Britain were formed, together with the names of the various periods of rock formation. Such a diagram is called a **geological column** and it is so useful that it is a good idea to make sure you are thoroughly familiar with it. Some people might even learn it off by heart.

20 What are the geological names for the rocks found in Iceland in Fig. 1.15?

21 By comparing Fig. 1.23 with the geological column in Fig. 1.24, work out which rocks were being formed at each of the past positions of Britain.

Fig. 1.24 The geological column: the youngest rocks are always shown at the top of the column

Millions of years ago	Earth movements	Geological period	Conditions and rocks in Britain
2		Pleistocene	Ice Age in Britain with warm periods
7		Pliocene	Warm climate; Crag rocks in East Anglia
26	Alpine	Miocene	No deposits in Britain
38		Oligocene	Warm shallow seas in south of England
54	Volcanoes in W. Scotland	Eocene	Nearly tropical; London Clay
136		Cretaceous	Chalk deposited; Atlantic ridge opens
195		Jurassic	Oxford Clays and limestones; warm
225		Triassic	Desert; sandstones, gypsum and salt
280		Permian	Desert; red sandstones, limestones
345	Volcanoes in C. Scotland; Cornish granite	Carboniferous	Tropical coast with swamps; coal
410		Devonian	Warm desert coastline; sandstones
440	Scottish granite	Silurian	Warm seas with coral; limestones
530	Welsh volcanoes	Ordovician	Warm seas; volcanoes, sandstones, shales
570		Cambrian	Cold at times; sea conditions
2800		Pre-Cambrian	Igneous and sedimentary rocks

The eruption of Krakatoa

Until the year 1883 few people had heard of Krakatoa. It did not appear in many atlases as its name would have been far longer than the size of an uninhabited tropical island would merit.

22 a) Find the position of Krakatoa between Java and Sumatra in your atlas.
b) Mark it on the earthquake map produced for Exercise 7.

In May 1883 a volcano on the island became active and erupted. People in Batavia, 160 km away, hired a steamboat to go and see the sight. As time went on the eruptions became more and more spectacular. By August people's curiosity was tinged with concern.

At ten o'clock on Monday, 27 August three tremendous explosions were heard and then came the loudest noise ever to be heard on earth. It was appallingly loud in Batavia. Five thousand km away to the west, across the Indian Ocean, a watchful coastguard on the island of Rodriguez noted the character of the sound and the time he heard it. The awful noise had taken four hours to travel the distance.

In his Christmas lecture to young people at the Royal Institute in 1901, Sir Robert Ball said: 'If Vesuvius were vigorous enough to emit a roar like Krakatoa, how great would be the consternation of the world. Such a report would be heard at Windsor ... and by the Tsar of all the Russias at Moscow. It would penetrate to the seclusion of the Sultan at Constantinople. Nansen would still have been within its reach when he was at his furthest north near the Pole.

'It would have extended to the sources of the Nile near the equator. It would have been heard at Mecca by pilgrims. It would have reached the ears of exiles in Siberia. No inhabitant of Persia would have been beyond its range, while passengers on half the liners crossing the Atlantic would have heard the mighty sound.'

23 a) On a world map, put the point of your compass on Vesuvius and draw a circle with a radius to represent 5000 km.
b) Mark on your map, with the help of your atlas, the places mentioned by Sir Robert Ball as within earshot.

☆24 Study the map of the plate boundaries of the world (Fig. 1.21) and try to explain why Krakatoa erupted. Fig. 1.17 should be helpful for this.

a)

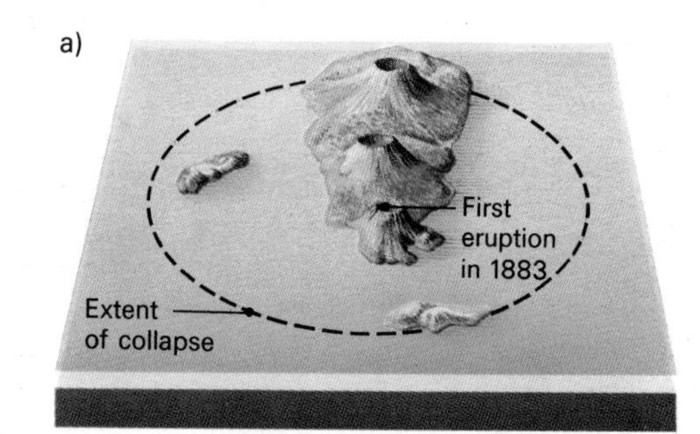

b)

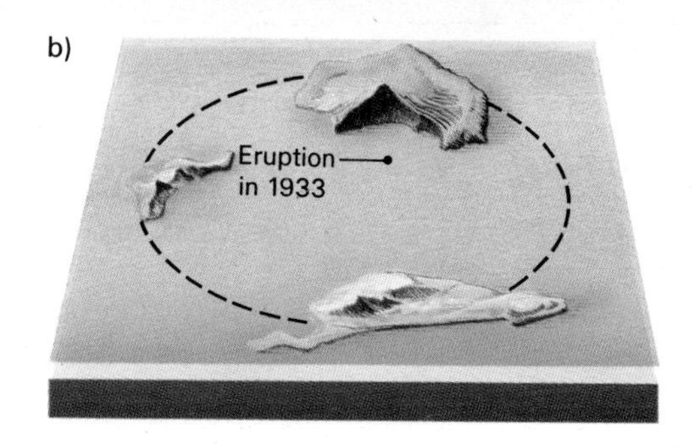

Fig. 1.26 Krakatoa: a) before and b) after the great eruptions of 1883

Fig. 1.25 A painting of an early stage in the eruption of Krakatoa

When Krakatoa erupted with such a stupendous noise, a vast glowing cloud of dust rose 80 km into the air (Fig. 1.25). Vast sea waves or tsunamis (Chapter 10, p. 114), one of them 60 metres high, swept over the low-lying coasts of Java and Sumatra, drowning 36 000 people.

When Krakatoa became visible again, it was found that two-thirds of the island (over 20 km²) had disappeared. At first it was thought that it had all been blown into the air. However a careful survey showed that it had collapsed into a vast hole left by the erupted magma. All that was left was a ring of little islands (Fig. 1.26).

Krakatoa and the jet stream

The ash blown into the air did not just fall into the sea around Krakatoa. Instead, men watched Krakatoa's dust and ashes start on a vast journey westwards round the world. Before the eruption no one knew of the existence of the easterly **jet stream** winds blowing at very high speeds 30 km above the earth's surface.

In the autumn of 1883 the newspapers were full of accounts of strange appearances in the sky from Ceylon, the West Indies, and other tropical places. The moon appeared green or blue, a very rare event, giving rise to the expression 'once in a blue moon'. In November the dust reached the skies above Britain and the evening sky was filled with a series of brilliant sunsets.

This was not the first time that volcanic eruptions had thrown up dust and ashes to tinge the British sky. The eruption of the Laki fissure in Iceland in early June 1783 led to a darkening of the sky and a rain of ash that killed three-quarters of all the sheep in Iceland. A little later its effects were noticed as far away as Selborne in Hampshire.

The Reverend Gilbert White, curate of Selborne, wrote in his journal: 'From 23 June to 20 July the sun, at noon, looked as black as a clouded moon, and shed a rust-coloured ferruginous light on the ground, and floors of rooms; but was particularly lurid and blood-coloured at rising and setting.'

25 What effect would you expect the layer of ash between the sun and the earth's surface to have on ground temperatures?

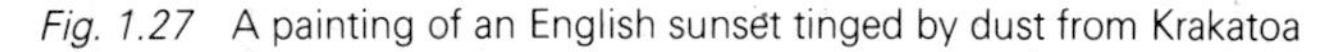

Fig. 1.27 A painting of an English sunset tinged by dust from Krakatoa

Summary exercises

26 Use the chapter to help you to explain the meanings of the following words:

Richter scale	mid-ocean ridge
seismograph	collision zone
epicentre	geological column
magma	jet stream

27 Find Hawaii, an island in the middle of the Pacific Ocean, in your atlas. Its position in the centre of the Pacific plate on Fig. 1.21 is something of a mystery. Can *you* think of any possible reasons for a volcano being there?

28 Make a list of volcanoes' names and try to find out from encyclopaedias and library books what sort of eruption they have had in the past and where they would fit on Fig. 1.13. Examples could be: Mt Hekla, Kilauea, Paricutin, Soufrière, Etna, Stromboli, Mt St Helens.

2
Shelter

Rocks and houses

Thousands of years ago early man sought shelter in rocks and caves; how much things have changed since then, or so we think! Yet we still use rocks and stones to build our houses. Look at the house in Fig. 2.1 and see just what rocks have been used to build it.

The chimney pots, ridge tiles, and bricks for the walls are made from clay. Notice the different colours of the bricks. The roof is made from slate while the gutters and drainpipes are made of cast iron. This is made from iron ore, which is another rock. Even the cement and mortar holding the house together are made from sand and chalk or limestone. In the picture the kerbstones are made of granite.

1 Draw a picture of your house or flat. Add labels to show the rocks used to make it.

Fig. 2.1 A house showing different types of rocks used in building

Some rocks were laid down as silt or sediment under the sea. These **sedimentary rocks** include:

i) clay which is formed from mud.
ii) sandstone which is formed from sand.
iii) limestone formed from the shells and skeletons of minute sea creatures and corals.

Other rocks come from the molten interior of the earth. If they have been laid down as lava on the surface of the earth they are called volcanic rocks. If they have been squeezed or intruded between other rocks, and then cooled down under the ground, they are called intrusive rocks. Both volcanic rocks and intrusive rocks were once hot and molten and are called **igneous rocks.**

When these igneous rocks passed close to sedimentary rocks, the intense heat and pressure changed the sedimentary rocks so that they became **metamorphic rocks**. Clays were squeezed so that they changed into slate, while limestone was changed into marble. Often minerals like copper, tin or even gold formed thick veins of pure metal in between the surrounding rocks.

Fig. 2.2 shows the different kinds of rock, while Fig. 2.3 shows where they are found in the British Isles.

2 Which rocks shown on Fig. 2.1 are sedimentary, igneous, or metamorphic?

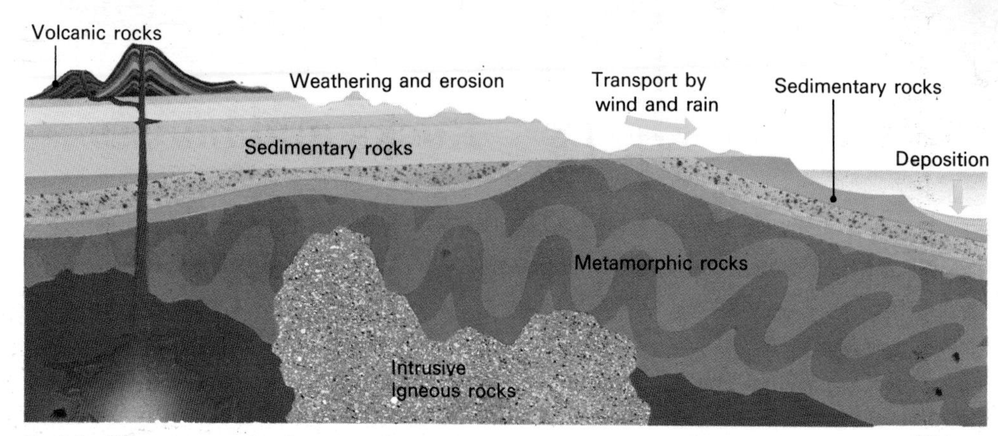

Fig. 2.2 Different types of rock

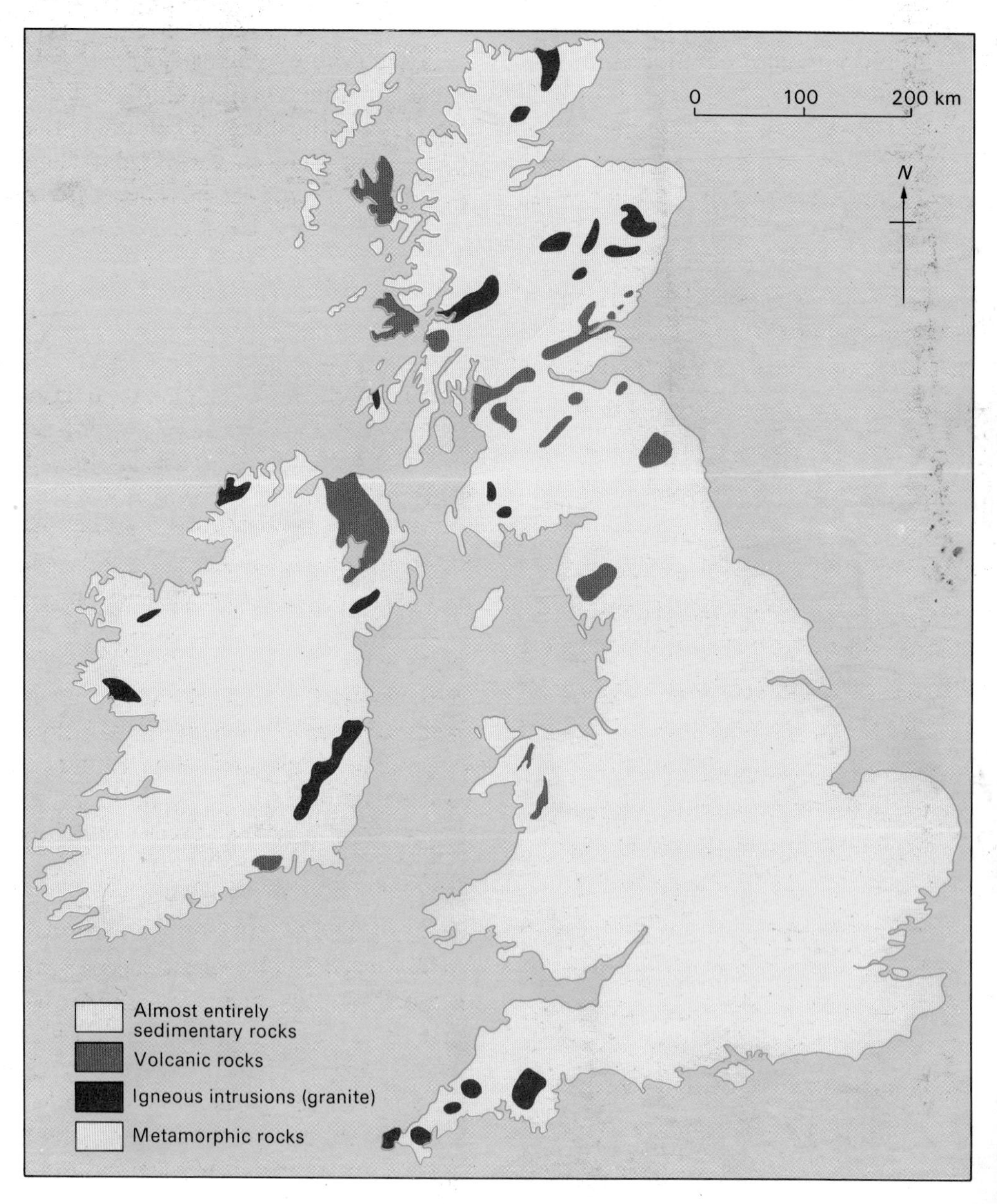

Fig. 2.3 Rock types in the British Isles

Granite

There are many parts of Britain from which the granite block used as a kerbstone in Fig. 2.1 may have come. A likely part is Dartmoor, a wild area of open moorland, bleak and treeless except in the valleys. The grey granite rock is often exposed in the upland areas of the moor as great granite **tors** (Fig. 2.4).

How did this rock come to be there? About 250 million years ago a great mass of molten rock squeezed itself in between layers of other rocks to form a huge blister. The molten rock never reached the surface. It pushed up the rocks above it and the intense heat and pressure melted the surrounding rocks in places. It was able to cool down very, very slowly to form a huge mass of granite as shown in Fig. 2.5.

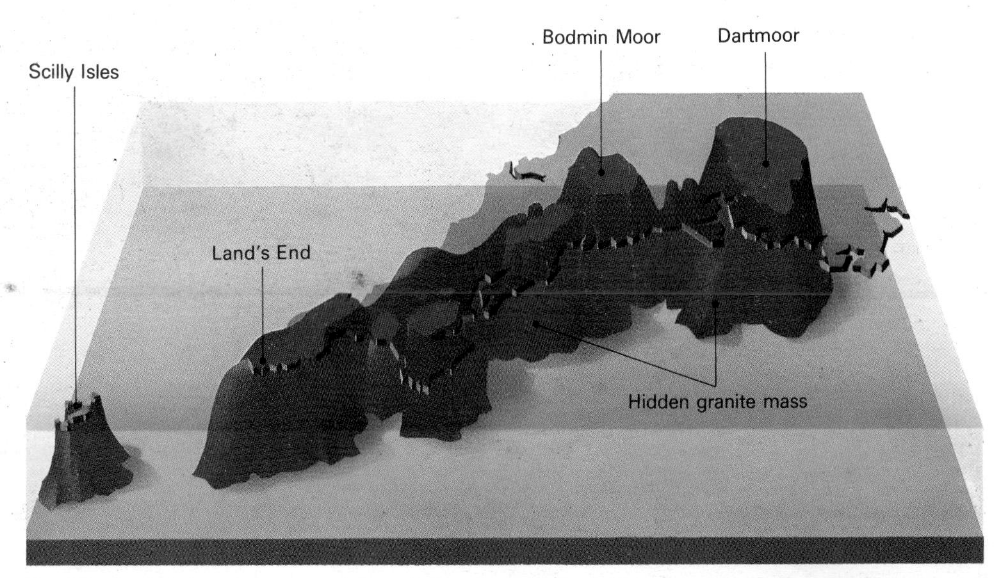

Fig. 2.5 The huge granite mass under south-west England

Fig. 2.4 Hound Tor on Dartmoor. Other tors can be seen on the skyline

The slowness of the cooling allowed the molten minerals in the rock to form large crystals of black shiny mica, glass-like quartz, and pink or white felspar (Fig. 2.6). These crystals make the rock very hard and in a cool, damp climate it resists the weather very well. This and the attractive mottled colour of the rock make it much sought after as a building material. It was often used for elaborately carved tombstones because it lasts so well.

The rock is also useful because it occurs naturally in blocks, as can be seen from Fig. 2.4. When it was formed deep down under layers of other rocks, the granite cooled under pressure. When the rocks above were washed away exposing the granite, this pressure was released and as a result great cracks or joints formed in the granite.

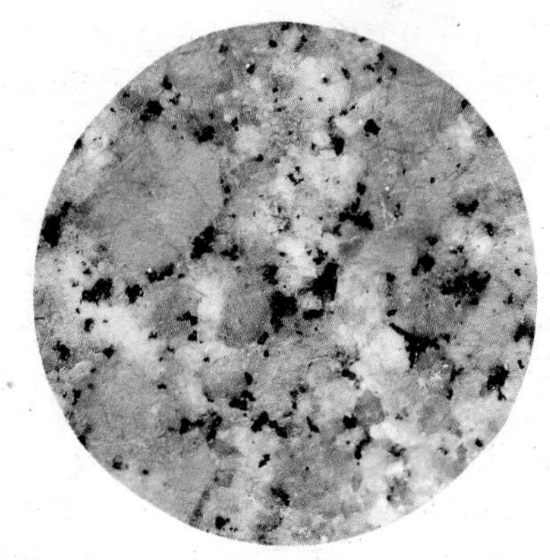

Fig. 2.6 Crystals in granite, shown actual size

There are very few towns where granite has not been used. It may be found in churches and graveyards, in shop-fronts, as kerbstones and bollards, and as road chippings.

3 Make a list of the examples you can find in your town where granite is used.

Clay

Fig. 2.8 Cattle grazing on clay countryside

Clay is soft and sticky when it is wet, but when it is dry it bakes very hard. It is made up of tiny grains with very little room in between them for air or water. Therefore, clay does not allow water to sink in from the surface and clay country has many streams, ponds, and lakes. Hills and slopes are mostly gentle, but because clay is so heavy to plough it is often left as grass with oak and ash trees standing in the hedgerows.

4 Which features of clay country can you see in Fig. 2.8?

Clay was once used as a very cheap building material in Devon, Kent and Hampshire, where it was known as 'cob'. As long as the clay was kept dry, it made quite a good wall for a house. The wall was usually built on a row of stones to stop damp creeping up from the ground. The clay was plastered over to stop rain washing it away and a roof of overhanging thatch protected the top.

Fig. 2.7 Main rocks used for brick-making

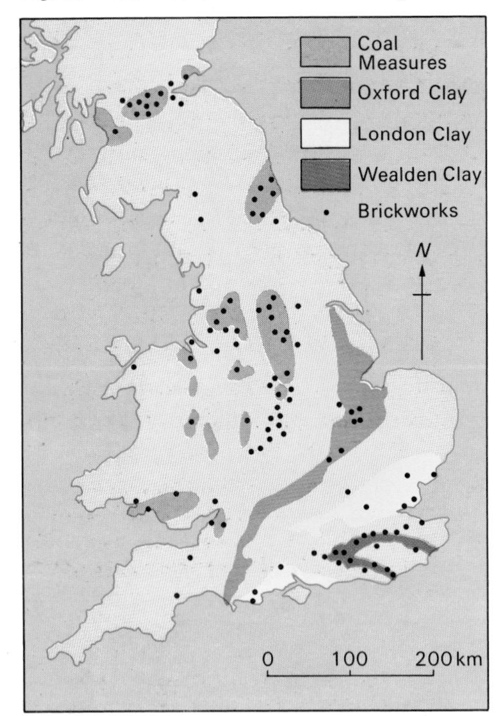

Clay is still used in building, but it is made into bricks first. There are more than a thousand brickworks scattered through Britain and Fig. 2.7 shows the main rocks used for brick-making.

The colour of the bricks depends on the type of clay used. London Clay makes yellow bricks. Stoke-on-Trent produces Staffordshire Blues, while Oxford Clay bricks are pink. The clay found between the seams of coal in the Coal Measures produces pink bricks, which can stand great heat (fire bricks), as well as dull red house bricks.

5 What colour bricks are found in your area? Find out if they come from a local brickworks or from further away?

Fig. 2.9 This brickworks has produced a very different landscape from Fig. 2.8

6 The sentences below explain the advantages of Oxford Clay for making bricks. Choose the correct ending for each sentence and then write out the complete sentences.

a) Oxford Clay is found in very thick layers over a wide area . . .

b) Oxford Clay contains its own carbon fuel . . .

c) Oxford Clay contains about 20 per cent moisture . . .

. . . so it requires less fuel to bake it into bricks.

. . . so it is easily pressed into brick shapes.

. . . so that a great number of bricks can be made at any brickfield.

Slate

The most important slate-quarrying area in Britain is in North Wales. It has produced the best slates in the world. The massive scars of the industry will remain for hundreds of years. Some people think such quarries are ugly, others find their huge cliffs have a strange beauty (Fig. 2.10).

Penrhyn quarry near Bethesda is the largest slate quarry in the world. It forms an oval-shaped amphitheatre, 2 km long and $\frac{3}{4}$ km wide. The series of galleries rise from near sea-level to over 400 m and the whole quarry is dominated by massive waste tips of reject slate.

Slate started its life many millions of years ago as clay or mudstone which was deposited under water. If you look at clay under a very powerful microscope, you will see that it is made up of millions of very tiny platelike minerals. One reason for clay being so soft and slippery is that these plates slide easily over each other. Normally the tiny plates lie at all sorts of angles to each other; the top left of Fig. 2.11 shows the plates as they were when the clay was first deposited about 500 million years ago.

Later, about 325 million years ago, huge earth movements squeezed the mud. Under this intense pressure all the plates came to lie at right-angles to the direction of the pressure put on them (Fig. 2.11). The soft clay became hard slate. This splits very easily in the direction in which the plates lie. It is called **slaty cleavage**.

Only a few taps with a hammer and chisel will make a huge block of slate split into slabs with smooth straight sides. These slabs can be split again and again, until they are the right thickness for roofing slate, that is about 3 mm.

Yet, if you want to cut slate in any other direction, you would need to use a very hard sharp saw, like the one you would use to cut very tough wood. Indeed, in the quarrying area of North Wales slate has been used for making many things which would be made of wood elsewhere: door posts, gate posts, even furniture.

Fig. 2.10 Penrhyn quarry from the south-east. Each terrace or gallery is 20 m high

7 Look out for buildings and houses with slate roofs in the area where you live. Some of them might have dates on them to say when they were built. Show where they are on a map of your area.

Slaty cleavage makes it very hard for water to seep through slate from one side to the other. Often a line of slates is built into the wall of a building about 30 cm above the ground as a damp course to stop rising damp entering the wall above.

8 a) Look at the outside of your home or school to see if there is this sort of damp course.
b) Draw a diagram to show why water cannot pass very easily from one side of a slate to the other.

9 Because slate is so hard and very few chemicals attack it, some benches and tables in laboratories have tops made of slate. Why should this make slate a good roofing stone?

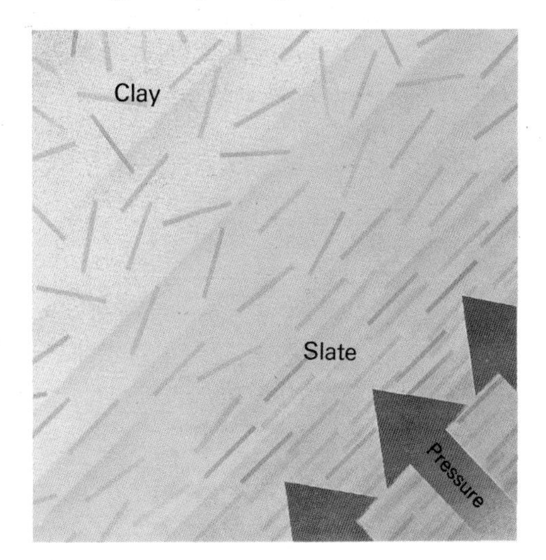

Fig. 2.11 How clay mineral plates are changed to slate

Limestone

Limestone is used in two main ways for building: as a building stone, for example in St Paul's Cathedral in London, and in cement.

Limestone also occurs in many people's homes in a very different way. If you live in an area of hard water, you will have seen what happens when water is boiled. 'Fur' appears as a coating on the inside of the kettle (Fig. 2.12). Hard water has a chemical called calcium carbonate dissolved in it from limestone. When the water is boiled, calcium carbonate is deposited inside the kettle.

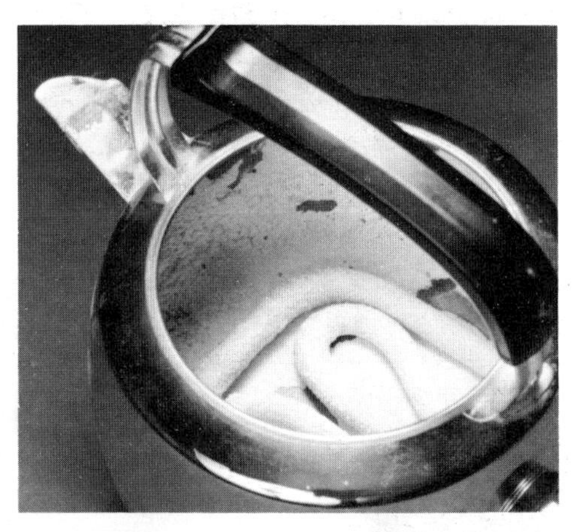

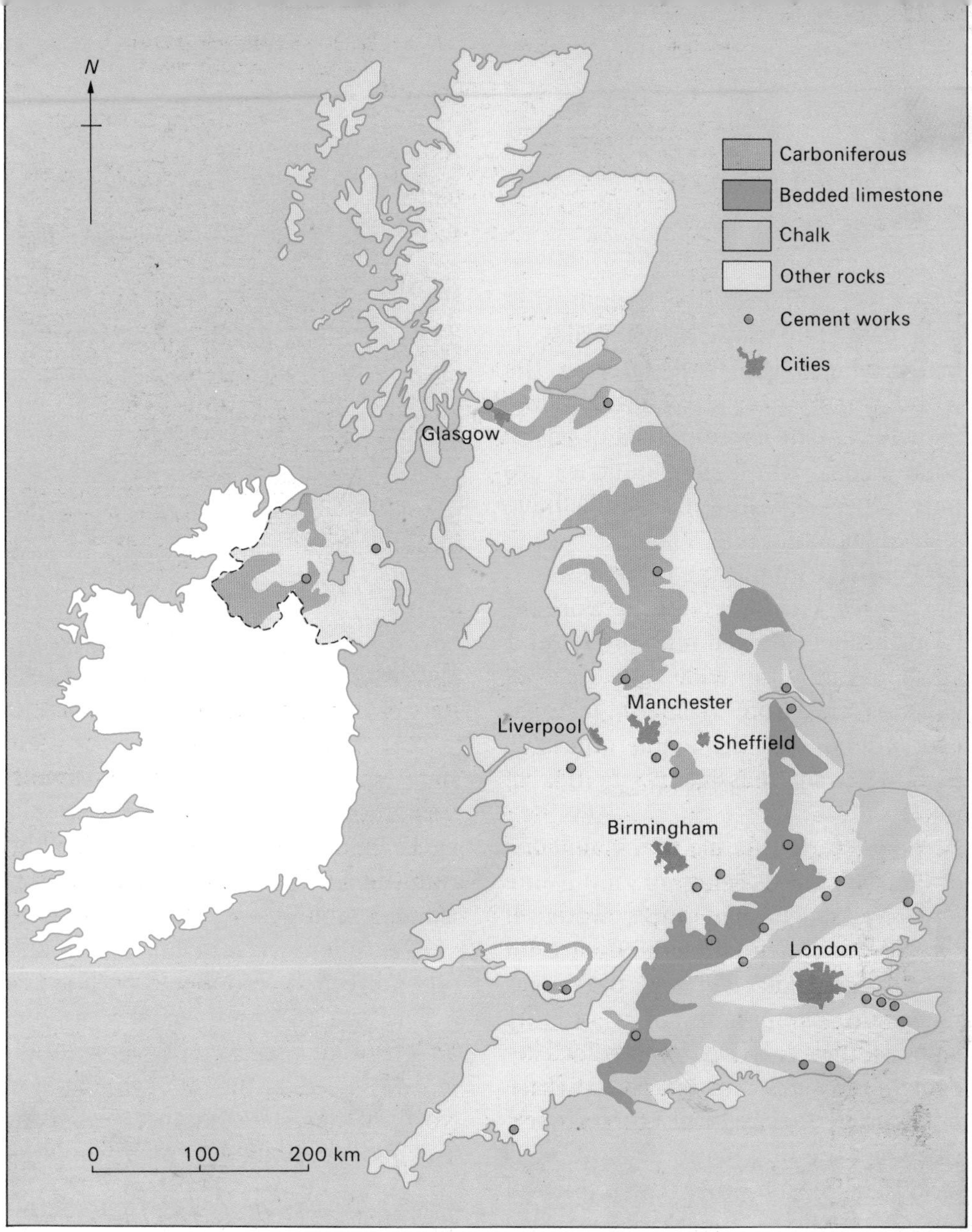

Fig. 2.12 (top left) Fur inside a kettle

Fig. 2.13 (bottom left) Coral under the sea

Fig. 2.14 (above) Main limestone areas of Britain and cement works

In a warm sea, helped by the shells and skeletons of small sea animals or corals, a deposit of limestone can form on the sea bed. A coral reef like that in Fig. 2.13 could become limestone in the future.

The best known limestone is probably Carboniferous Limestone, where the calcium carbonate forms little crystals. Then there are the bedded limestones, which run in a belt from Dorset through the Cotswold hills and up through Northamptonshire. They provide the best building stone in England.

Chalk is the purest limestone; it is

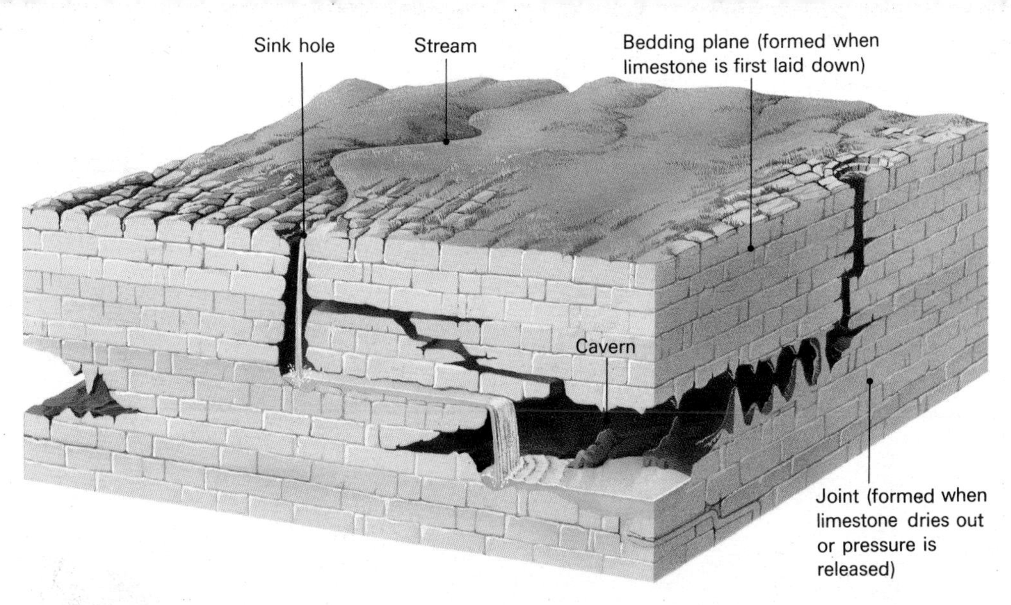

Fig. 2.15 Joints and bedding planes in Carboniferous limestone

Fig. 2.16 A deep cave in France: stalactites hang from the roof, stalagmites mount from the floor

not often used as a building stone because it tends to be soft and dissolves easily in rain water, but it is very useful for making cement. Fig. 2.14 shows where the main limestone rocks are found in Britain.

Limestone is a good building stone mainly because it is a freestone and can be cut or split in any direction, unlike slate. It is laid down in beds with vertical cracks or joints rather like those found in granite. These joints allow water to sink through the rock.

10 a) Find some samples of different types of rock such as limestone, chalk, sandstone, granite, slate, and clay.
b) Carve a hollow in the top of each and tip a little water into the hollows. Which rocks allow water to sink or permeate through and which are impermeable?

The crystals in Carboniferous Limestone leave few gaps to allow water through, so most of the water moves along the joints and bedding planes (Fig. 2.15). As the water moves through the limestone, it dissolves a little of the rock on either side of the cracks, making them wider. In time these cracks may open up into caves or caverns as shown in Fig. 2.16.

There are very few streams flowing over the surface of limestone country because so much water sinks underground. Streams that start off flowing over other kinds of rocks often sink down swallow holes when they reach the limestone, as in Fig. 2.17. They may then flow underground until they reach the surface again as a 'resurgence' when the rock changes.

Fig. 2.17 Block diagram of an area of limestone country

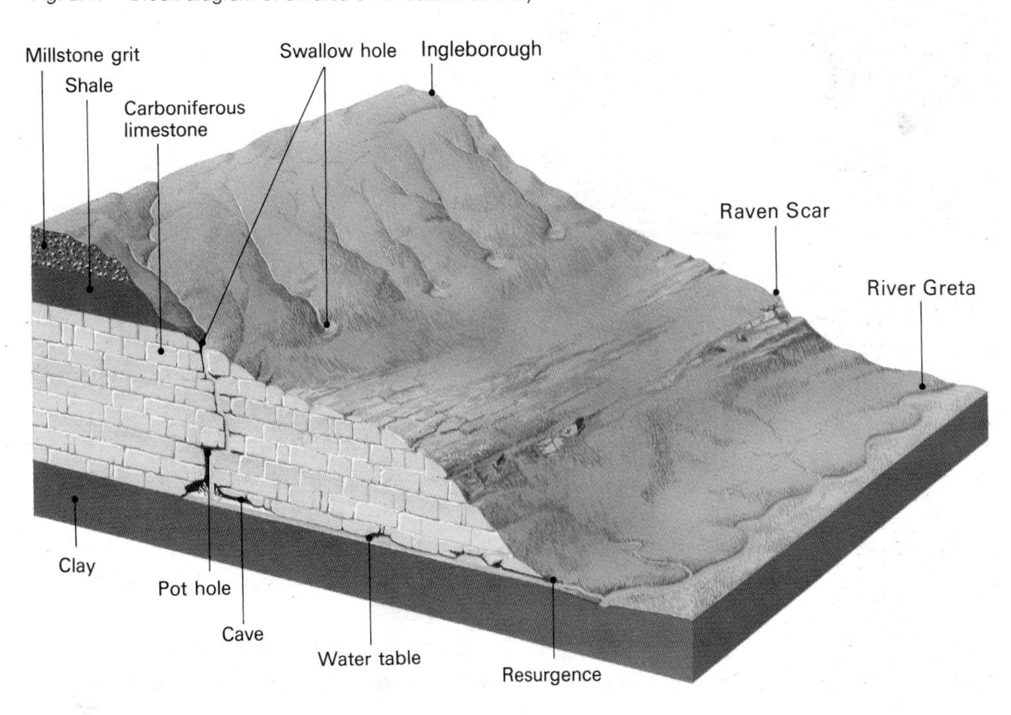

Most **erosion** (or wearing away) of limestone happens underground. In the end the roof of a cave may collapse and a gorge can be formed. The walls of the cave form the sides of a narrow valley like that at Cheddar Gorge (Fig. 2.18).

The natural erosion of limestone is a slow process, but a much more rapid form of erosion is shown in Fig. 2.19. Britain makes 18 million tonnes of cement every year. Most limestone areas have their cement works; Fig. 2.14 shows how many there are.

For every tonne of cement manufactured, more than one and a half tonnes of rock must be extracted and transported to the works. Chalk is preferred as it is the purest and softest limestone. About four-fifths of the raw material for cement is chalk or limestone; the rest is largely clay with a little gypsum (the raw material for plaster of Paris) to help the cement to set.

11 Look at Fig. 2.14. Make a list of the type of limestone used at each of the cement works shown.

The quarries used today for digging out chalk or limestone are often backfilled or filled with rubbish. Then the top soil is put back and the land is used again for farming. Not all quarries can be filled in again. Quarries are an example of how people are the most powerful agent of erosion.

Fig. 2.18 Cheddar Gorge: an impressive sight visited by many thousands of people each year

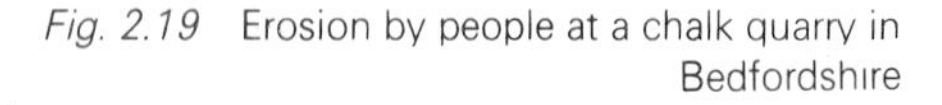

Fig. 2.19 Erosion by people at a chalk quarry in Bedfordshire

Gravel

Look carefully at Fig. 2.20. It shows the River Findhorn flowing from the right of the picture and curving round away from the camera. On either side of the river itself is a flat part of the valley which, because it is liable to flood at times, is called the **floodplain**.

On either side of the floodplain is a low wide **river terrace**; this is made up of gravel laid down by the river when it was flowing at a higher level. Above this terrace is another older terrace formed in the same way from gravel.

Fig. 2.21 shows the terraces on either side of the River Thames in London. The oldest terrace is the highest one. It was laid down when the river was flowing about 50 metres above its present level. In these ancient gravels are found the remains of elephant and rhinoceros, as well as flint tools. At Swanscombe in Kent, the skull of the earliest known man in Britain was discovered.

Below this is the Taplow terrace in which the remains of woolly mammoth have been found, showing that the climate at the time the gravels were laid down was much colder. These bones are also found in the lowest terrace, the Floodplain terrace.

Besides being useful for providing gravel to make concrete, the terraces are good building sites, giving a firm foundation as well as being well drained.

Fig. 2.20 Gravel terraces above the floodplain of the River Findhorn in Scotland

12 Find out if there are any river terraces in your area and what has been found in them. What are they being used for today?

☆**13** How high above present sea-level was the River Thames flowing when it laid down the gravel between Clapham and Westminster which now forms:
a) the highest terrace on Fig. 2.21?
b) the Taplow terrace?
c) the Floodplain terrace?

Fig. 2.21 Cross-section of the Thames valley showing the river terraces

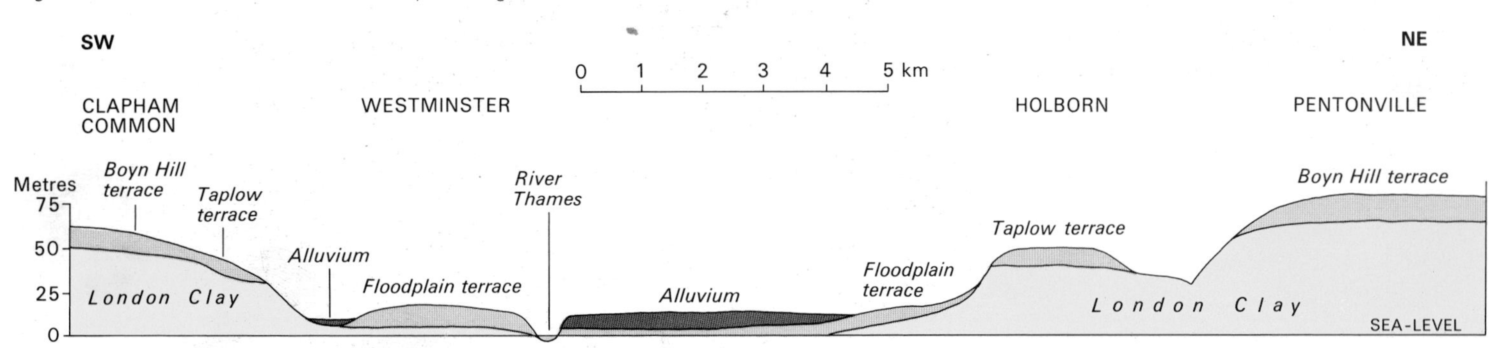

The geological map

A geological map shows the rocks which underlie the countryside. Part of a geological map is shown in Fig. 2.24. Running from north to south is the River Itchen; its floodplain is filled with **alluvium**, the silt and debris washed down by the river over thousands of years.

The flat land of the floodplain makes it ideal for playing-fields, if rather wet at times. It does not provide a good foundation for buildings, as can be seen in the photograph of the Bishop's Palace in Winchester (Fig. 2.22). The crooked sills of the windows show that the foundations have slipped.

The Cathedral (482293) was built on the gravel of the river terrace running along the western edge of the floodplain; later it was extended eastwards onto the floodplain and here the walls can be seen to lean.

Quite steep slopes mark the western edge of the river terrace, where it meets the Middle Chalk hillside. The Upper Chalk gives still steeper slopes, and the deep railway cutting (473271) has white chalk sides.

Fig. 2.22 Note the crooked window sills

14 Draw a sketch-section along the grid line 275 between Oliver's Battery (460275) and Deacon Hill (505275). Mark on St Catherine's Hill (483275) and the course of the River Itchen (476275).

Winchester has grown mainly on the chalk ridge running from west to east on either side of the Itchen valley. The ridge is called Teg Down (460298) on the west and St Giles's Hill (490293) on the east. The chalk was laid down originally as flat beds. It has since been folded upwards by earth movements (Fig. 2.23). A huge arch was formed over what is now called Bar End (490285) between St Catherine's Hill in the south and St Giles's Hill in the north.

The folding of the chalk weakened the top of the arch and made it crack. This was easily washed away so that what was once an upfold or anticline is now the Chilcomb Valley with Lower Chalk as its floor (Fig. 2.23).

15 Draw a sketch-section from east to west along northing 285 on Fig. 2.24. Label the Upper Chalk, the Middle Chalk, the river terrace, the floodplain, and the Lower Chalk.

Although chalk is the purest form of limestone, it often contains lumps of an impurity called **flint**. This is hard and does not dissolve in water. Flint is often used as a building stone in chalk areas and was one of the stones used by ancient man for making his stone tools. Capping the tops of hills in the area shown on the geological map is a layer of **clay with flints**.

Some geologists think that the clay with flints is the remains of impurities left over when the rest of the chalk was dissolved or washed away by rain. They think that clay with flints is mainly found on the tops of hills because these are more exposed to the weather.

Fig. 2.23 The formation of the Chilcomb valley

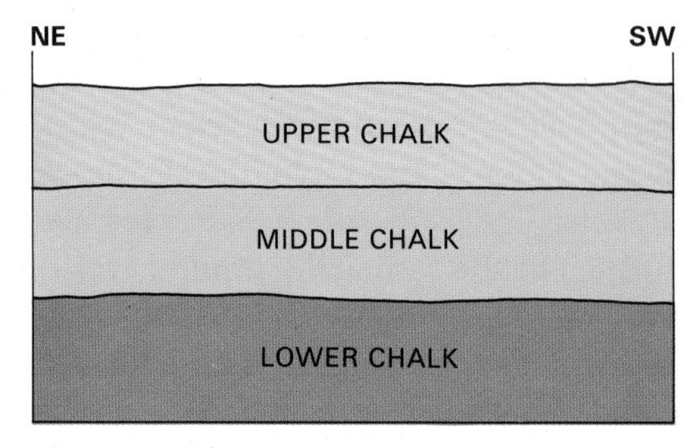

a) The chalk lies in flat layers.

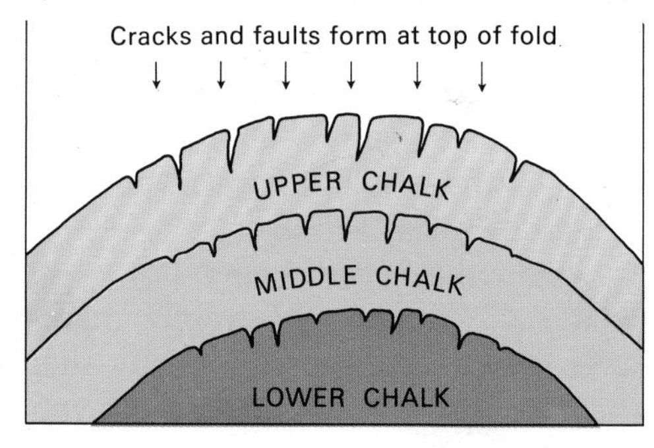

b) Pressure from the south folds the rocks, weakening them.

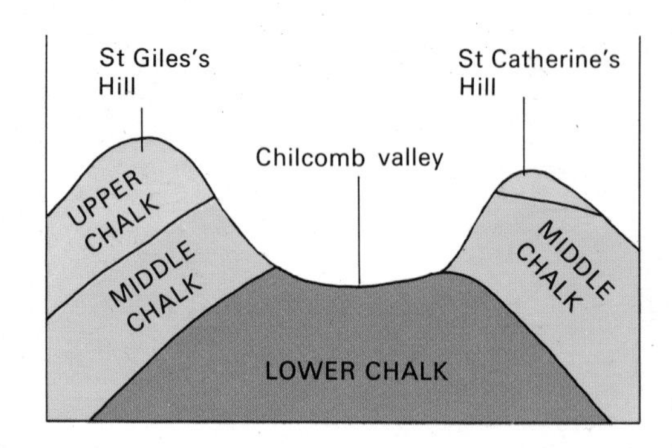

c) The weakened rocks are easily washed away.

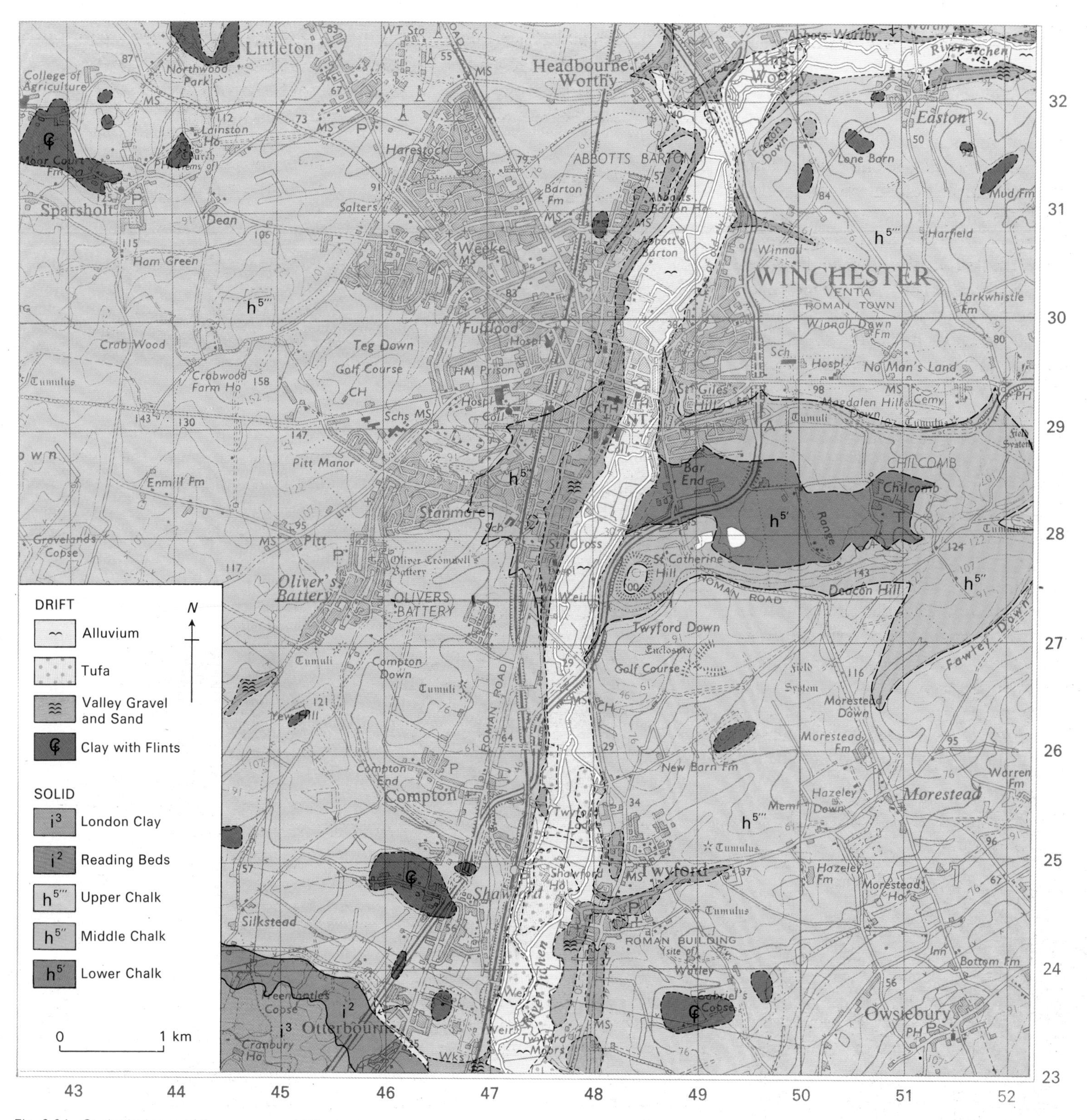

Fig. 2.24 Geological map of the area around Winchester

Other geologists argue that chalk does not contain minerals which can make clay. They point out that the flints are not knobbly and lumpy like those found in chalk. They are often rounded flint pebbles, like pebbles that have been rolled along by water on a beach or in a stream. The flints are stained a rusty brown throughout. They think that because of this the clay with flints is the remains of younger rocks that were laid down on top of the chalk and have since been weathered away.

Fig. 2.26 A dry valley in chalk country: the Devil's Dyke near Brighton

16 Fig. 2.25 shows two geologists arguing. Professor Perry Klein thinks clay with flints comes from the weathered chalk below; Dr Crystal O'Grapher thinks that it comes from the younger rocks above.
a) Make a larger copy of the drawing and fill in their argument in speech balloons above their heads.
b) Who do you think is right?

On either side of the main Itchen valley are smaller valleys leading into it. These have no rivers in them and so are called **dry valleys**, for example, at 485265. A dry valley is shown in Fig. 2.26.

Chalk allows water to sink or permeate through it, so it is called a **permeable rock**. If there was nothing underneath the chalk, the water could go right through the rock and out the other side. However, underneath the chalk there is a layer of clay which does not allow water to sink through. The lower levels of the chalk are full of water and in chalk country wells are dug to supply water.

These wells fill with water up to the level of the **water table**. In chalk the water table curves to follow the ups and downs of the land surface. In Fig. 2.27 and in the area of chalk shown on the geological map, the water table lies below the level of the valley floors and so there are no rivers in these valleys.

Although there are no rivers in dry valleys now, there were streams in them once; the streams eroded away the chalk to form valleys when the water table was higher.

☆**17** a) Trace the pattern of contours from the geological map (east of easting 48 and south of northing 28).
b) Mark on your tracing where you think the rivers once flowed.

Fig. 2.25 Two geologists arguing

Fig. 2.27 A block diagram showing a dry valley in chalk

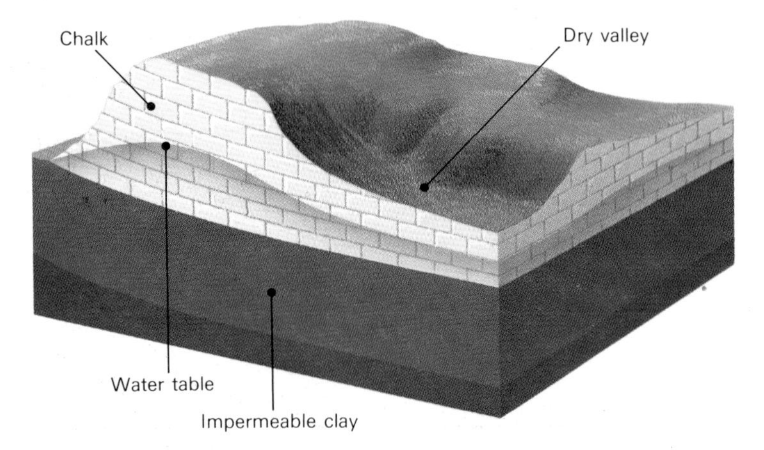

Protection from the weather

At the start of this chapter we said that early man looked for shelter in caves and hollows in the rocks. In all the buildings we have looked at rocks are still used to provide protection from the weather, including the wind. In Britain the wind blows most frequently from the south-west: this is the **prevailing wind**.

18 a) Look at Fig. 2.29 which is a diagram showing the points of the compass: it is called a **wind rose**. Every day for a week at the same time observe from which direction the wind is blowing. Then shade in one of the spaces on a copy of the diagram.
b) At the end of the week the pattern of winds will be clear. Is the south-west wind the prevailing wind?

19 In part of Fig. 2.28 the buildings have been shaded to show how exposed they are to wind and driving rain from the south-west. On a copy of the picture use the key to help you complete the shading. Show which parts of the buildings are most exposed to south-westerly winds and which are least exposed.

We shall see in Chapter 3 (p. 41) that buildings can raise the temperature of the air in towns, but they can also have the effect of cutting out sunlight. This can be important in many ways.

If buildings are too close together, one building will completely cut out direct sunshine from another. This will have the effect of increasing the need for electric light and heating in the shaded building. The effect on plant growth in gardens can also be severe.

Buildings are usually designed to protect people from the effect of the wind, but they can have quite the opposite effect. This is especially true of

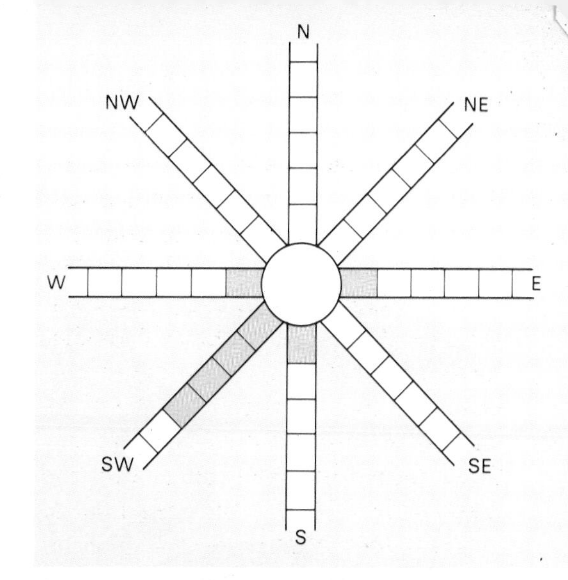

Fig. 2.29 Using a wind rose

Fig. 2.28 Part of a town showing exposure to the south-west wind

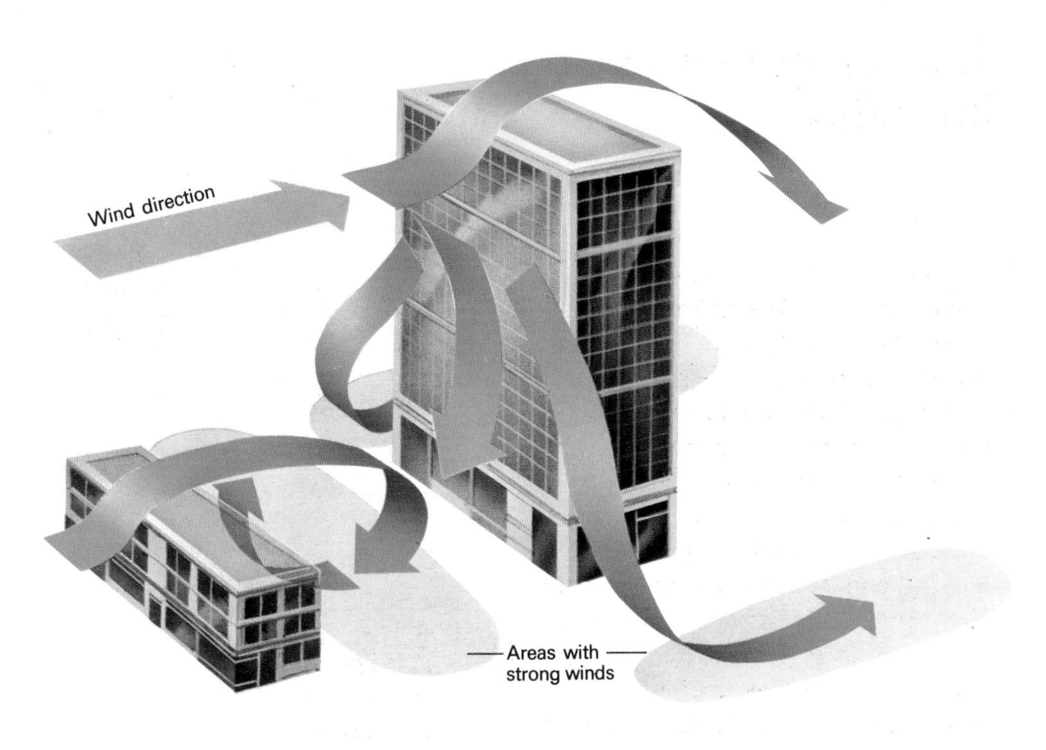

Fig. 2.30 Wind blowing round buildings

Fig. 2.31 The roof was built to protect shoppers

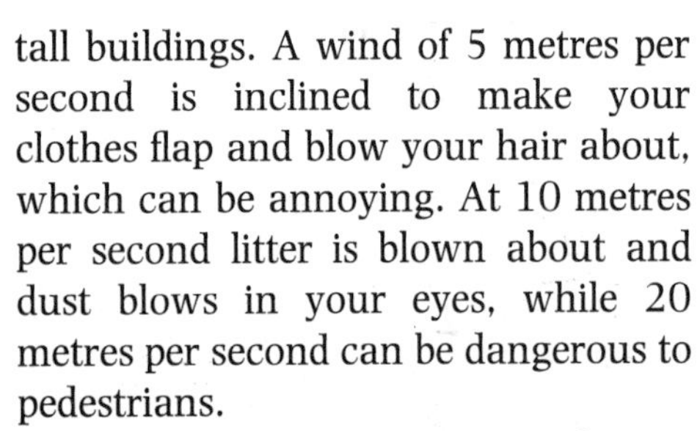

tall buildings. A wind of 5 metres per second is inclined to make your clothes flap and blow your hair about, which can be annoying. At 10 metres per second litter is blown about and dust blows in your eyes, while 20 metres per second can be dangerous to pedestrians.

In Fig. 2.30 you can see that at low levels the wind blowing from the left curves over the low building and then downwards. At ground level on the downwind side of this building the wind is actually blowing in the opposite direction from when it started. In the gap between the two buildings there will be a spiral eddy of wind, like a corkscrew lying on its side.

At higher levels the wind hits only the taller building. Some flows up and over the top. Some goes down to join the spiralling eddies below, while some is deflected down and round the tall building. This jet of air flows four times as fast and with sixteen times the force of the normal wind. It can be very dangerous for old people walking round the building.

Fig. 2.31 shows one example of a solution that had to be adopted in a shopping precinct at the foot of a tall building in Croydon, south of London.

20 What will happen to fumes from a chimney placed on top of the lower building in Fig. 2.30? On a copy of the diagram draw in the path of these fumes.

21 a) Can you find examples in your home area or school which suffer from high speed winds round tall buildings?
b) Do you know of any rows of trees in country areas planted as a wind break or **shelter belt** to protect crops from the prevailing wind?

Summary exercises

22 Use the last chapter to help you explain the meanings of the following words:

sedimentary rocks	river terrace
igneous rocks	alluvium
tor	clay with flints
slaty cleavage	prevailing wind
floodplain	shelter belt

23 Draw a map of Britain showing the places where traditional types of stone have been used for building. You could illustrate your map with sketches of houses of different types found in different parts of the country.

3 Warmth

Temperature and humidity

'... temperatures tomorrow will be around 20°C ...' We all know that the **thermometer** in Fig. 3.1 will tell us how hot the weather is. However, very few people know how hot the weather will be when the weather man says it will be 20°C tomorrow.

Anders Celsius was a Swedish scientist. He divided his thermometer into 100 equal parts between the point where water boiled and the point where water froze. The temperature scale most often used today is named the **Celsius scale** after him, though we usually call it **centigrade.**

1. Use the picture in Fig. 3.1 to help you to say when you would go out of doors:
   a) without a coat.
   b) without a jumper or jacket.
   c) wearing a swim-suit.

2. a) Hold up one of your fingers; does it feel hot, warm or cold?
   b) Now lick your finger to make it wet; does it feel hotter or colder?
   c) Now blow on your finger; how does it feel now?

Exercise 2 shows that how hot or cold the air feels depends on: (i) how damp or humid the air is, (ii) how fast the wind is blowing.

When you licked your finger and blew on it, it felt cold. The water evaporated from your finger into the air; this means it changed from a liquid to a vapour. To change like this, water needs energy. It uses up energy in the form of heat from your finger. This means your finger felt cold because of the heat it lost.

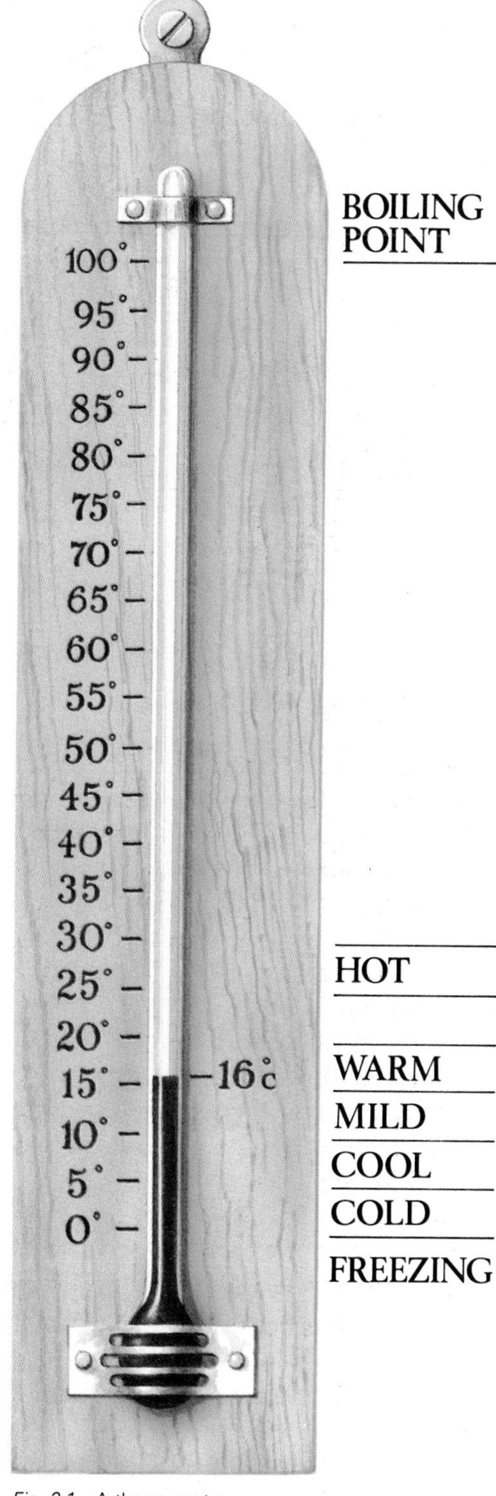

Fig. 3.1 A thermometer

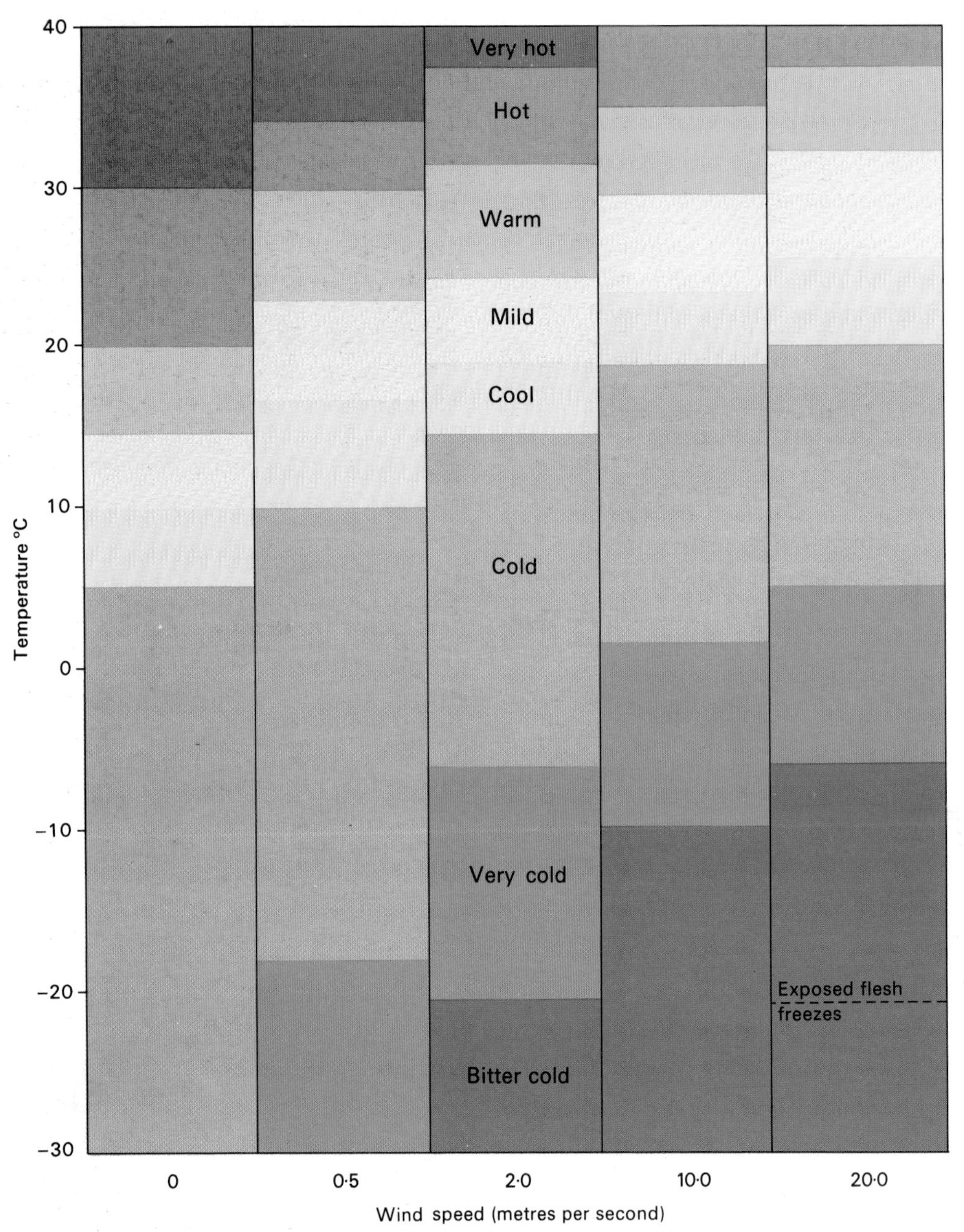

Fig. 3.2 How wind makes the air feel colder

Temperature °C	Humidity % (dampness of air)	Conditions
45	10	Too hot (heatstroke)
35	90	
22	50	
5	10	
8	90	
−5	50	

Fig. 3.3 Table to describe the conditions in a room according to the comfort chart (Fig. 3.5)

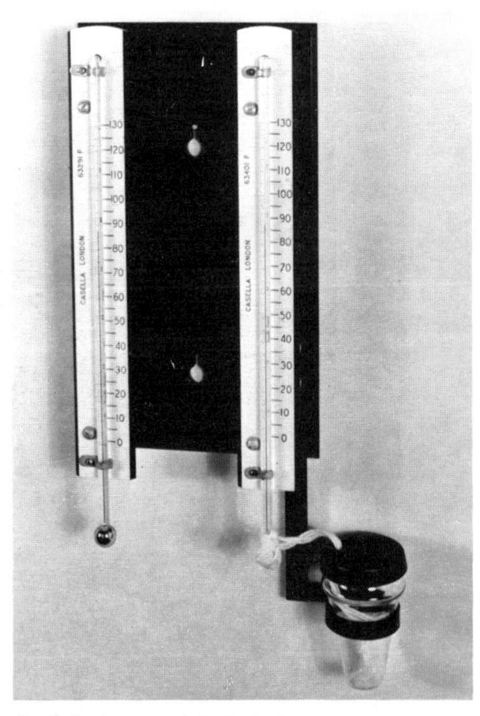

Fig. 3.4 A wet and dry bulb thermometer

Fig. 3.2 shows that wind makes us think that the weather is cooler than it really is. It also shows what a danger strong winds can be when the weather is below freezing.

Fig. 3.5 shows the effect of the dampness of the air on how comfortable a room feels. The higher up the diagram, the hotter the room is. The further to the right, the damper or more humid the air is.

3 Copy the table in Fig. 3.3. Then fill in the last column to describe what the room would feel like.

The instrument that measures how damp the air is is called a **hygrometer**.

The simplest form of hygrometer is a single human hair. One end is attached to a wall with a pin and a small weight is hung from the other end. The more humid the air, the longer the hair stretches.

A more usual kind of hygrometer is shown in Fig. 3.4. This consists of two thermometers. One thermometer has its bulb wrapped in muslin which is kept damp with water from a little bottle; the other thermometer has its bulb kept dry.

The more humid the air is, the less water will evaporate from the wet muslin. Therefore the temperature of the wet bulb thermometer will be closer to that of the dry bulb.

4 Fig. 3.7 shows the temperature and humidity in the middle of Kew Gardens at lunch time. If you went there on a visit, during which months would you want to have a picnic out of doors?

5 Readings of humidity are often taken:
a) in a greenhouse.
b) in art galleries.
c) in computer centres.
Explain in each case why you think this is done.

☆**6** Fig. 3.1 shows the lowest temperature allowed under the Offices, Shops and Railway Premises Act in places where people work. This is 16°C.
a) Do you think it is (i) too low, (ii) just right, (iii) too high?
b) What do you think ought to be the highest humidity allowed by the Act?
c) What is the temperature and/or humidity in your room at the moment?

Fig. 3.6 A sunny day in Kew Gardens near London

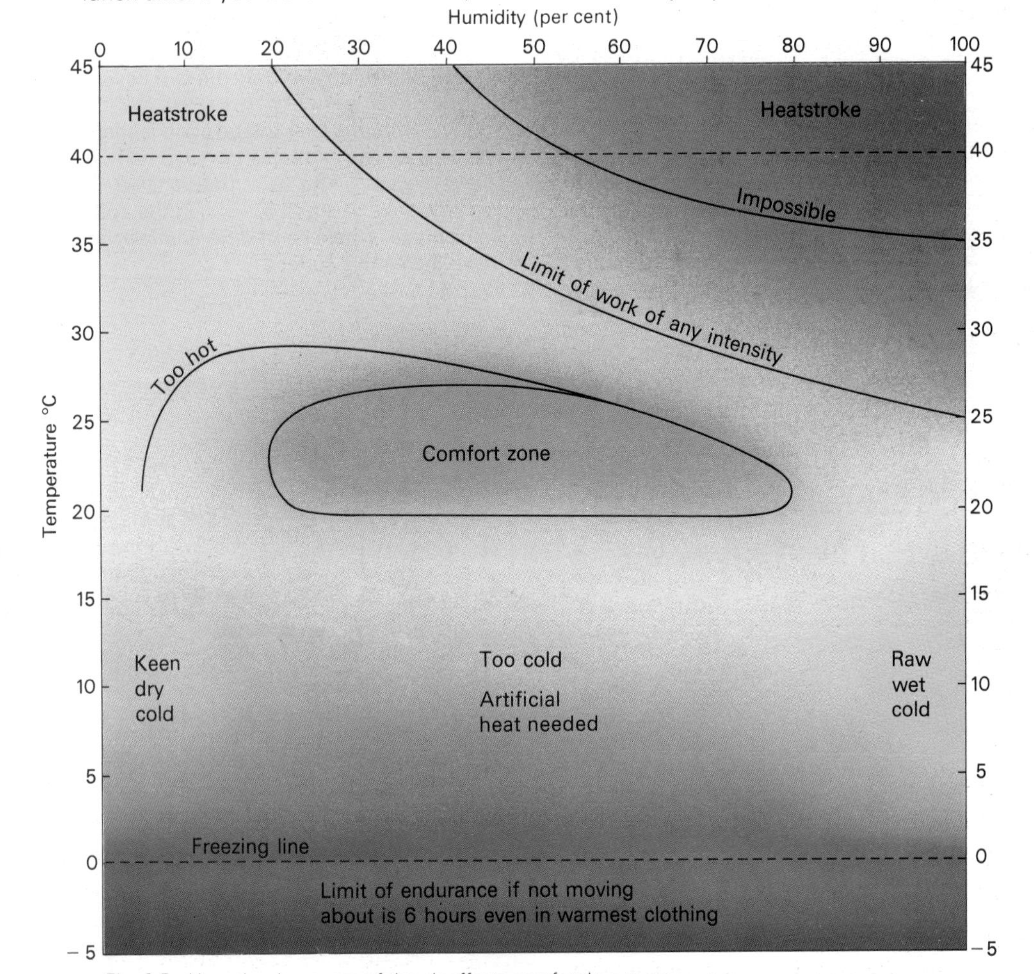

Fig. 3.5 How the dampness of the air affects comfort in a room

Month	Maximum temperature °C	Humidity %
January	8·8	84
February	10·3	83
March	13·3	80
April	15·8	74
May	19·2	73
June	23·2	72
July	24·3	72
August	25·3	75
September	22·2	78
October	17·7	84
November	11·8	86
December	9·7	85

Fig. 3.7 Temperature and humidity in Kew Gardens

The seasons

Fig. 3.9 A school globe on a tilted stand

The earth spins round once every 24 hours. The part of the earth that is facing the sun will be in daylight. Later the same part will be in darkness, because it is facing away from the sun. We can show this effect on a school globe when light shines onto one side of the globe, for example from a window or a lamp (Fig. 3.9).

A long time ago people thought that the sun went round the earth. Yet really it is the earth that is moving round the sun. It travels round the sun once every $364\frac{1}{4}$ days and this causes our seasons (Fig. 3.8).

If the earth spins round once every 24 hours, why does daylight last longer in summer than in winter? Fig. 3.9 shows the school globe with its axis, the line on which it spins, tilted at an angle.

Because the earth is tilted, in July the northern part of the earth is nearer the sun's heat and light (Fig. 3.8). The result is longer daylight hours and warmer temperatures in the northern hemisphere. This explains why temperatures are higher in Britain during the summer (Fig. 3.7).

7 Write out the passage below. Then use the list of missing words to fill in the gaps.

Missing words:

south	sets
six	winter
tilted	colder
spins	axis
north	sun

The earth ...... round the sun once a year. It also spins on its ...... once a day. In July the north pole is ...... towards the sun. Places ...... of the equator get more than 12 hours of daylight and less than 12 hours of night. Near the north pole in July the sun never ....

...... months later the position is the other way round. The earth has travelled half way round the .... The ...... pole is tilted towards the sun and it is ...... in the northern hemisphere, where short days and long nights make the earth's surface ....

Fig. 3.8 How the earth's journey round the sun causes the seasons

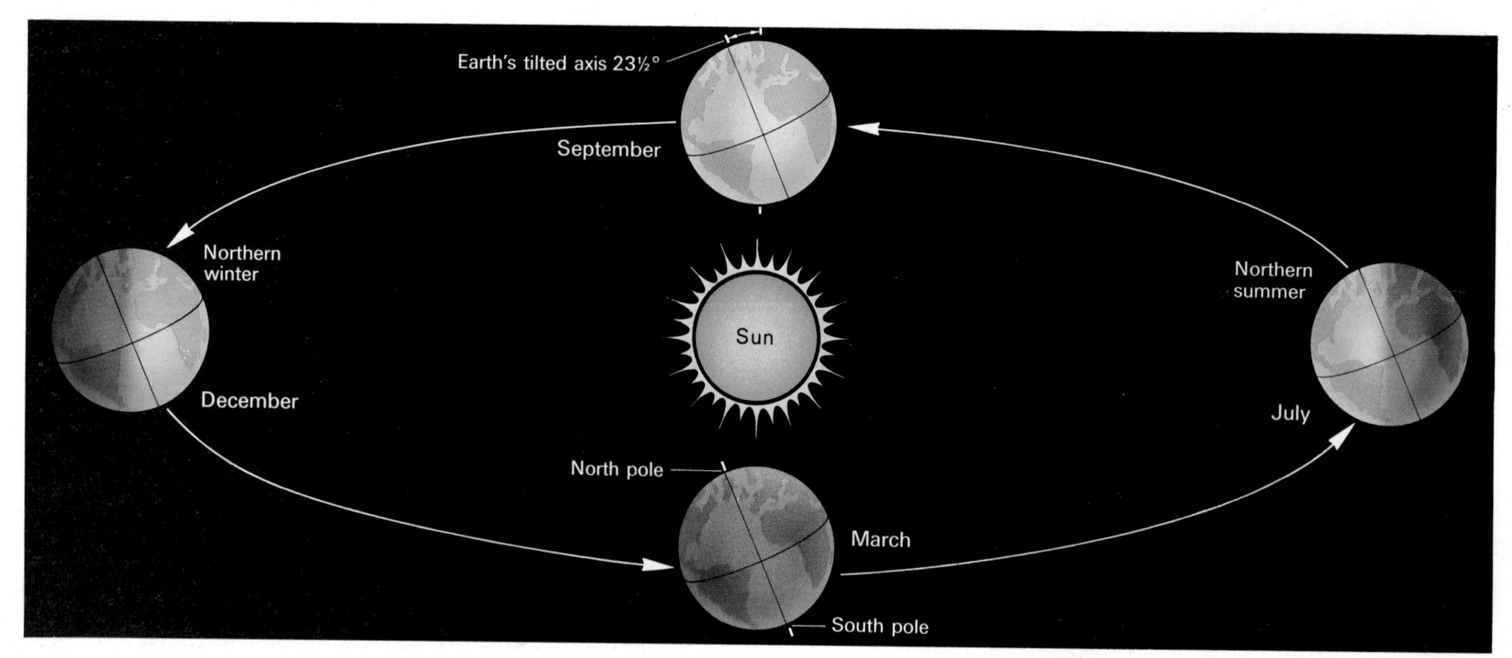

How the sun heats the earth

The following list explains what happens to 100 units of heat from the sun:
24 units are reflected back to space by clouds.
6 units are scattered in the sky, giving it its blue colour.
17 units are absorbed by the atmosphere.
4 units are reflected back to space from the earth's surface.
49 units heat up the earth's surface.

8 a) How many of the 100 units are reflected back to space?
b) How many are left to heat the earth and the air?

9 Use the text to help you fill in a copy of Fig. 3.10 to show the correct number of units of heat for each arrow.

10 A photograph like Fig. 3.11 is made from reflected light. Which parts are:
a) light reflected from clouds?
b) light reflected from the ground?
c) blue scattered light?

Fig. 3.11 A photograph of the earth taken from the Apollo 17 spaceship

Fig. 3.10 What happens to 100 units of heat from the sun

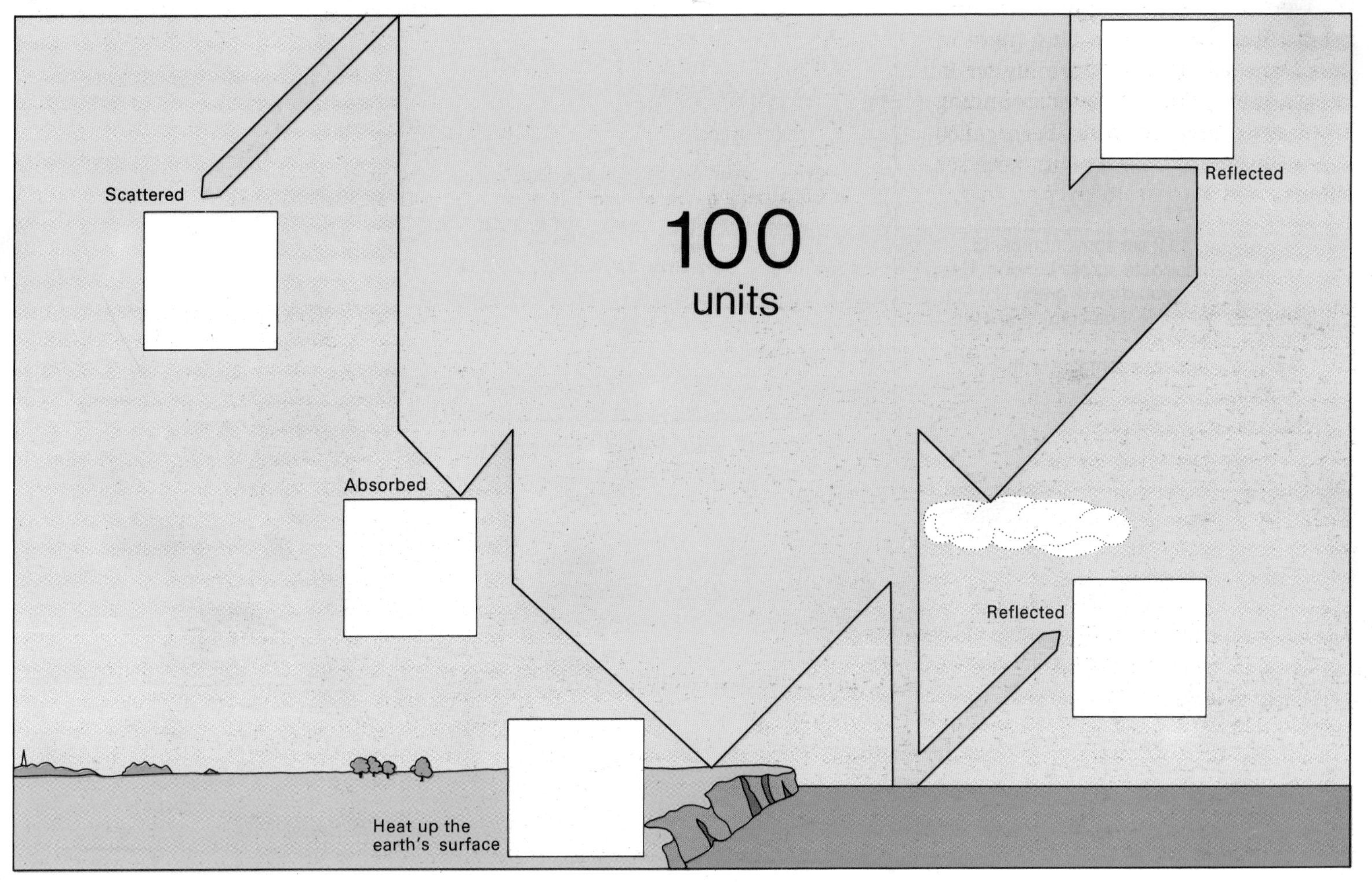

Urban heat islands

Fig. 3.14 shows a landrover to which thermometers have been fitted. One night in October the landrover was driven across London along the route shown on the map in Fig. 3.12. The graph in Fig. 3.13 shows how the temperature varied on the way.

11 a) On the night in October where is the coldest part of London?
b) Where is the warmest part of London?
c) Why is it warmer on the first journey than it is on the return journey?

Fig. 3.12 shows the pattern of temperatures in London on a night in June. The temperatures are higher in the city than in the surrounding country. This pattern has been called an **urban heat island.** Can you see why?

12 a) On Fig. 3.12 are some numbered points on the route across London. Use these to help you draw a graph like Fig. 3.13 for the June temperatures shown on Fig. 3.12.
b) How does this compare with Fig. 3.13?

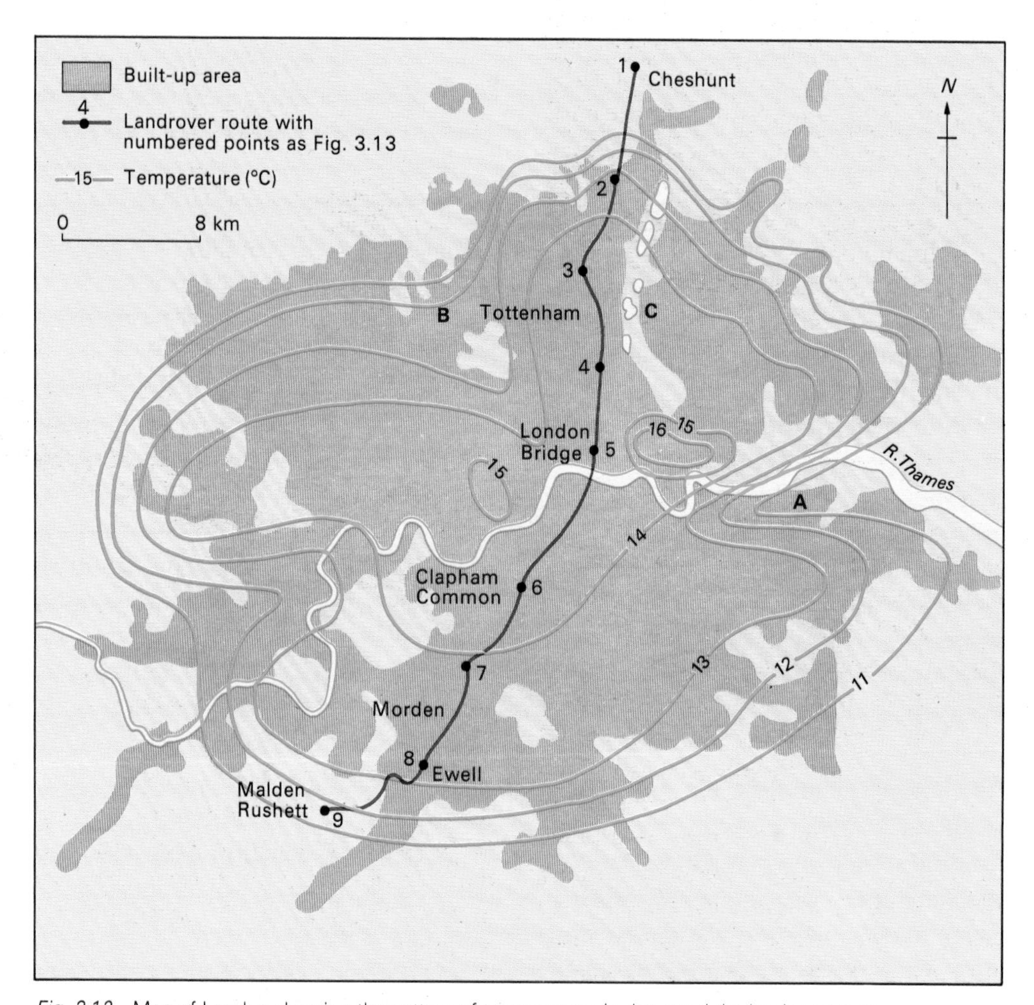

Fig. 3.12 Map of London showing the pattern of temperatures in June and the landrover route

Fig. 3.13 Temperatures recorded by the landrover on its journeys across London one October night and a cross-section of the route shown in Fig. 3.12

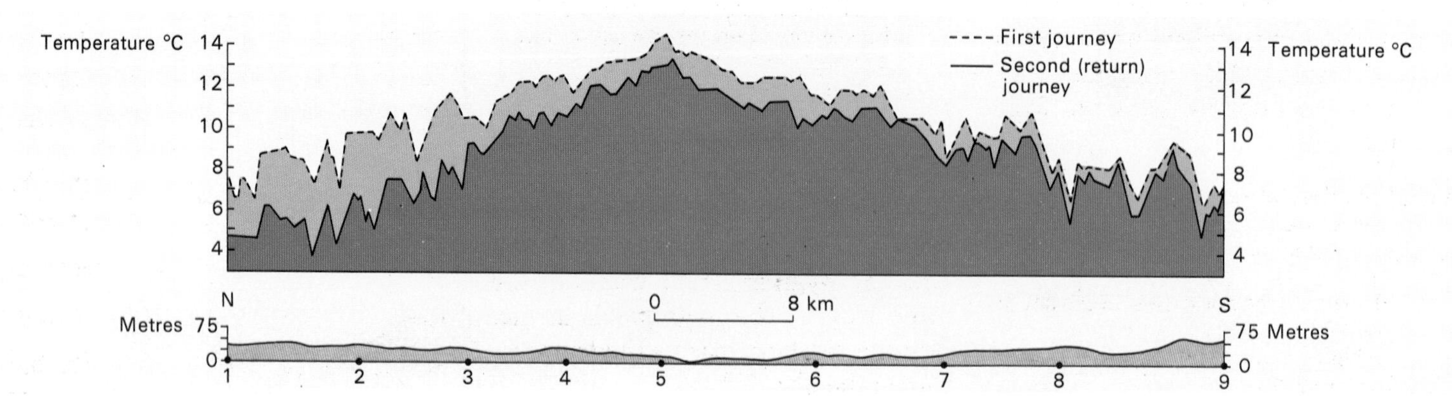

Fig. 3.14 The white box on the landrover holds thermometers

The following are some of the reasons why cities are warmer than the country around them:

i) Buildings often allow some of their heat to escape outside.

ii) Towns have many flat surfaces which reflect and absorb heat.

iii) Bricks, concrete and tarmac all store up heat during the day which can be released at night.

iv) Drains and sewers take away water in the towns which would lie about in the countryside, gradually drying up and evaporating. In the country this evaporation again uses up heat, which lowers the temperature of the countryside (remember the finger-licking experiment, p. 35).

v) The haze and dust that hang in the air above cities allow through heat from the sun, while reflecting back much of the heat radiating up from the city.

13 Look again at Fig. 3.12. What might have caused the temperature to be:
a) lower in the Thames valley at point A?
b) lower to the north-west of the city at point B?
c) higher in the industrial zone of the Lea valley at C?

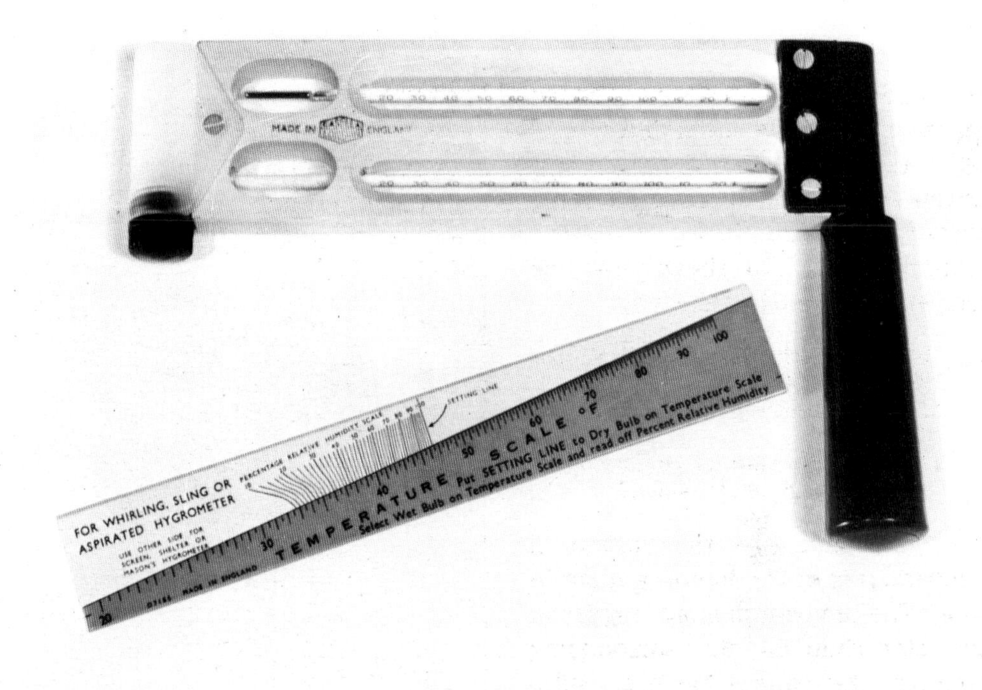

Fig. 3.15 The slide rule is used to work out humidity from the temperatures on the whirling hygrometer

Fig. 3.16 shows how much warmer temperatures are in the centre of many cities than in their surroundings; it also shows the population of each city.

☆**14** a) Draw a scattergraph with population on the horizontal axis and temperature differences on the vertical axis to illustrate the information in Fig. 3.16.
b) Is it true to say that the larger cities are much warmer than their surroundings?

Fig. 3.15 shows a very special wet and dry bulb thermometer. It is attached to a frame rather like a noiseless football rattle and is called a **whirling hygrometer**. It works in the same way as the hygrometer in Fig. 3.4; the dry bulb tells us the temperature of the air.

Fig. 3.16 Temperature differences between cities and their surroundings compared with city sizes

City	Temperature differences °C	Population
Berlin	10·2	3 137 000
London	10·0	7 168 000
Vienna	8·1	1 859 000
Sheffield	8·0	507 000
Malmo	7·4	451 000
Munich	7·1	1 337 000
Karlsruhe	7·1	261 000
Uppsala	6·5	136 000
Utrecht	6·0	463 000
Lund	5·7	35 000
Reading	4·2	133 000

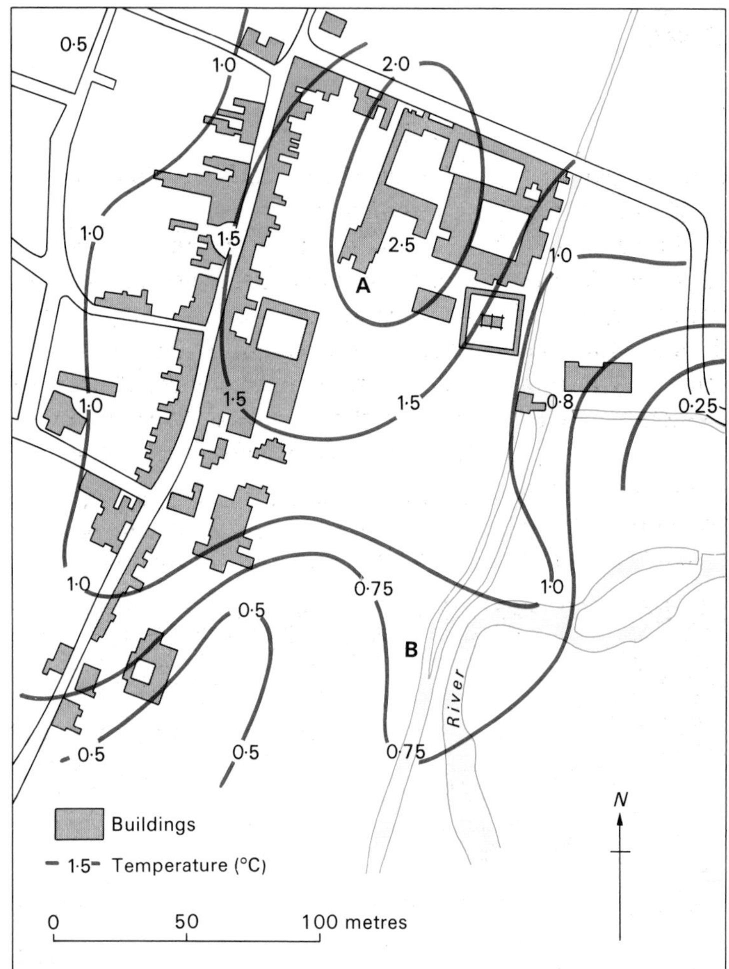

Fig. 3.17 Temperature in February at the edge of a town

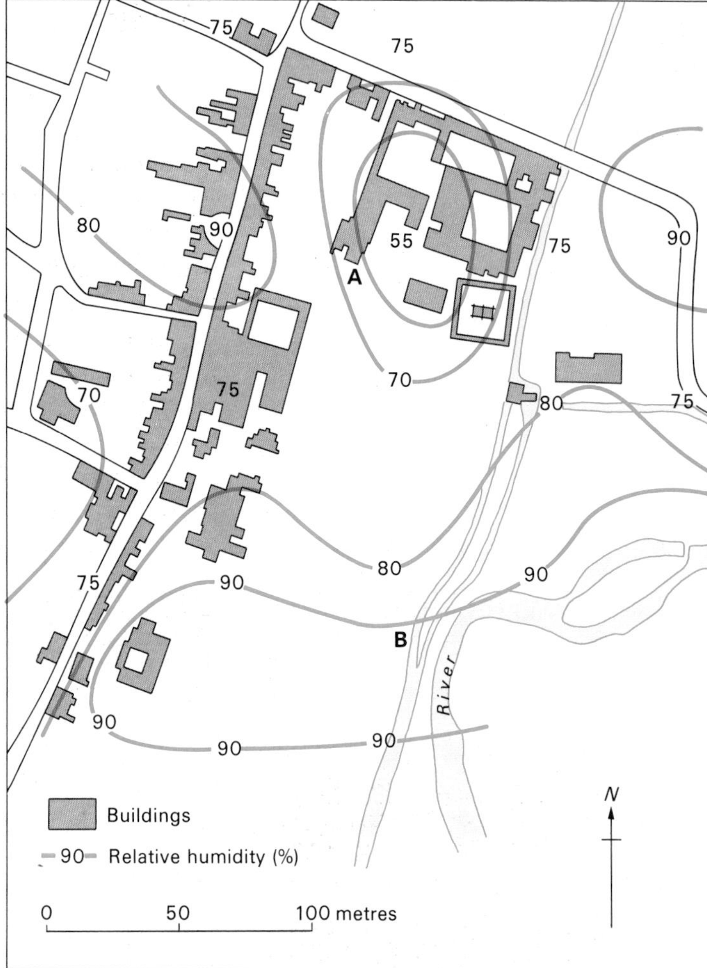

Fig. 3.18 Humidity in February at the edge of a town

When we stand and revolve the whirling hygrometer for a minute, the water will evaporate from the muslin on the wet bulb. The difference between the two temperatures will tell us how humid the air is.

We can use a whirling hygrometer in different places very easily. For example, we can see how temperature changes from one side of a town to another as in Fig. 3.13 or we can draw maps of temperature as in Fig. 3.12. You could use one to see if your town has a heat island. It is best to do the survey on a calm day or night so that the wind does not spoil the pattern.

15 Figs. 3.17 and 3.18 show the result of such a survey. What would you say is the effect on both temperature and humidity of:
a) buildings at point A?
b) water at point B?

Summary exercises

16 Use the last chapter to help you explain the meanings of the following words:

thermometer	hygrometer
centigrade	winter
humidity	summer

17 Use a diary to find the date when:
a) the daylight lasts longest in midsummer—the summer solstice.
b) the days and nights are the same length—the spring equinox and the autumn equinox.
c) the nights are longest in midwinter.

4

Water

Water in the air

Sometimes, after a summer shower, we can see steam rising from a wet road or a playground. The sun's heat turns water into vapour which rises into the air. This water vapour is invisible. The colder air above the road turns the water vapour back into small water droplets, which we see as steam.

The process of turning water into vapour is called **evaporation**. When wet or humid air is cooled, the vapour turns back into water: it **condenses** on to any surface.

This is why bathroom or kitchen walls will often run with damp, and why cars left outside at night will often have a thin film of water over them and will sometimes be very difficult to start in the morning. At such times, the grass will be covered with **dew**.

1 Look at the photographs in Fig. 4.1. Make a list of the types of condensation and add any others you can think of.

Fig. 4.1 Some different types of condensation

Clouds are formed of condensed water vapour. The water vapour condenses as small water droplets around bits of dust, ice, and salt floating in the air. The salt is sea spray which has been blown inland.

2 On a copy of Fig. 4.2 add labels chosen from the key to describe each of the arrows.

Air which is warmer than its surroundings tends to rise. Smoke from a fire or a chimney and thunderclouds in the sky both show this happening.

Very often we cannot see the hot air rising because the air is invisible, but we can see the heat shimmering above a hot road or roof, or gliders spiralling upwards on a rising current of air.

Normally dry air cools by about 1°C for every 100 metres it rises as in Fig. 4.3. Eventually if it continues to rise, it will reach a low enough temperature for the water vapour in the air to condense into clouds. This temperature is called the **dew point** or **condensation level.**

3 On Fig. 4.3 what is the temperature of the rising air at:
a) ground level?
b) 1000 metres?
c) the condensation level?
d) 2000 metres?
e) 3000 metres?

The height at which water will condense varies according to how wet or humid the air is and how warm it is. When the clouds are low it means that the air is either very humid or very cool.

4 Look out of the window. Is the cloud base high or low? Can you see any signs of rising air currents?

Fig. 4.2 The water cycle

Fig. 4.3 Temperature changes of rising dry air

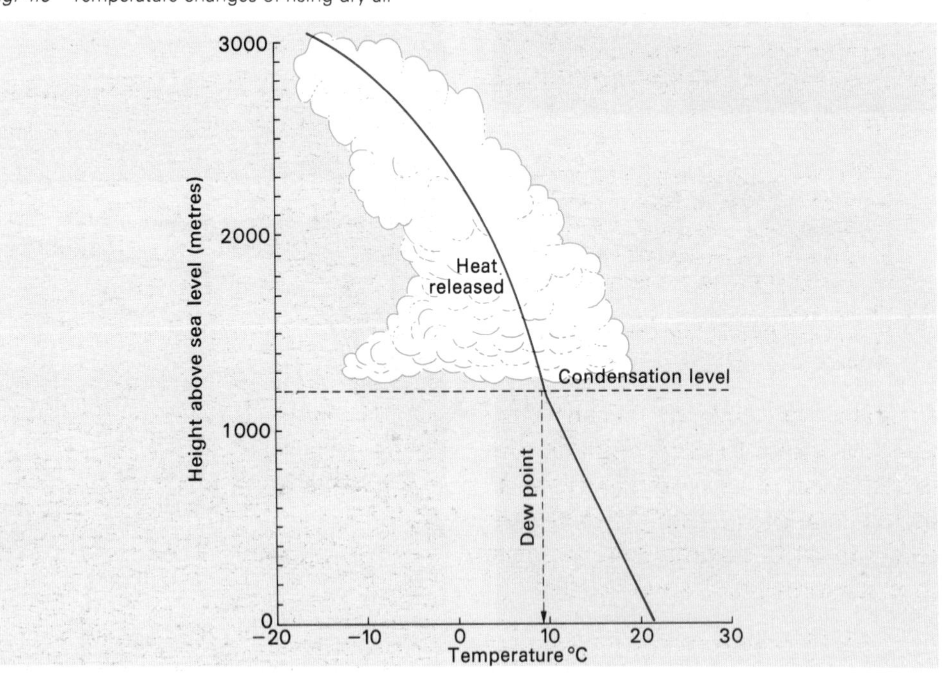

5 Use Figs. 4.2 and 4.3 to help you write meanings for the following words:
a) evaporation.
b) condensation.
c) cloud base.
d) dew point.
e) surface run-off.

Fig. 4.4 shows a balloon floating in the air at 1000 metres above a weather station. It has a thermometer to measure how warm the air is and a radio which sends the air temperature down to the ground; this is called a **radiosonde** balloon. Its thermometer is reading 4°C. At the weather station on the ground the air temperature is 10°C.

Warm air from the airfield is rising. When it left the ground it was 15°C. It cools at a rate of 1°C for every 100 metres, so it is 5°C when it reaches the same height as the balloon. The rising air is still warmer than the surrounding air so it will carry on rising.

Air that tends to rise is called **unstable air**. It will continue to rise until it has cooled down to the same temperature as the air around it. The air is now **stable** and will tend to sink downwards.

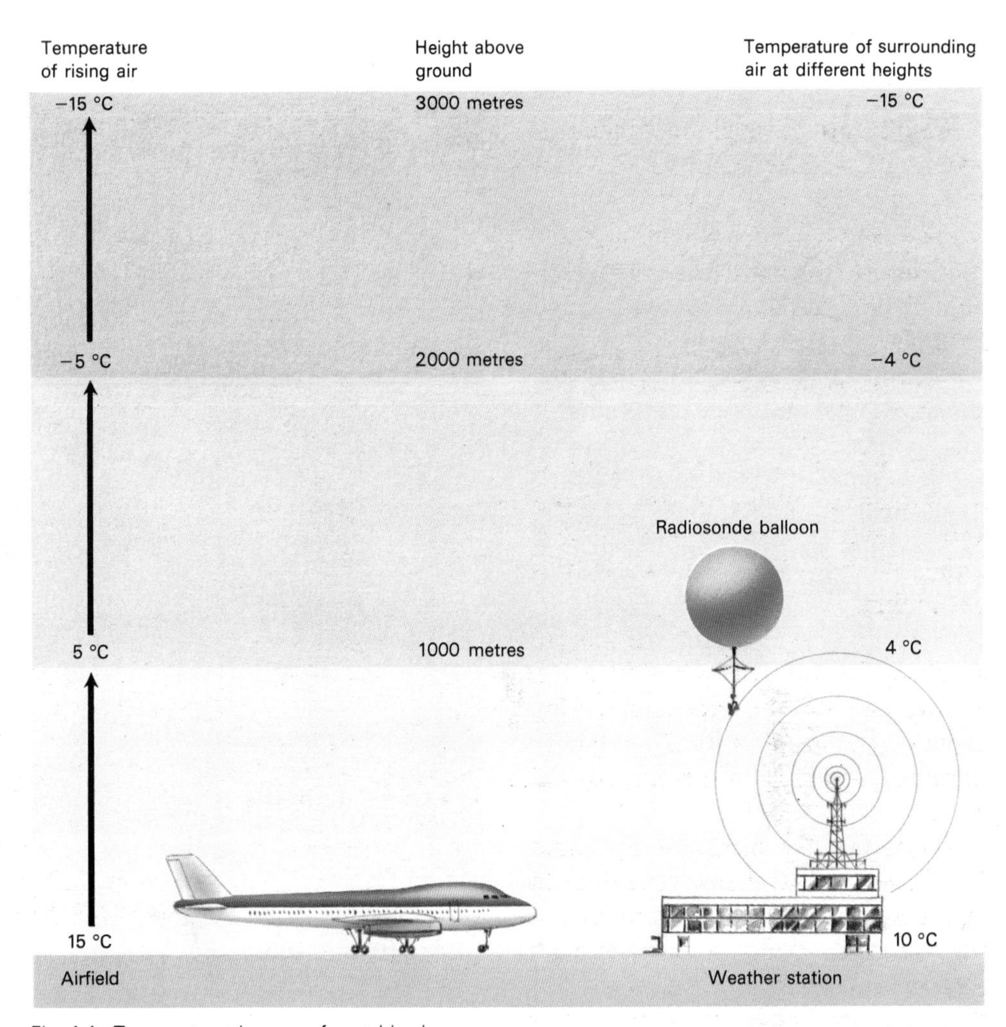

Fig. 4.4 Temperature changes of unstable air

6 a) Draw a graph showing the temperatures recorded by the balloon as it rises. Use the figures in Fig. 4.4 to help you.
b) Use a different coloured line on the same graph to show the temperatures of the rising air.
c) At what height does the air become stable?

Stable air tends to give us clear weather during the day. Fog and mist tend to occur at night and to stay until quite late in the morning. Unstable air tends to give rise to rain or even thunderstorms, hail, snow or sleet.

Fig. 4.5 Unstable air rising in a storm cloud and causing rain

Rainfall

There are three main ways in which air can be forced to rise and become unstable:

i) Warm damp air drifting over the sea may meet hills or mountains; it will be forced to rise as shown in Fig. 4.7 and cool to the dew point at which clouds will form. If the air is stable, then a thin layer of cloud will form with a flat base.

If the air is unstable, then the air will continue to rise with very thick cloud forming and rain or snow may fall. This leads to the air passing over the mountain losing moisture.

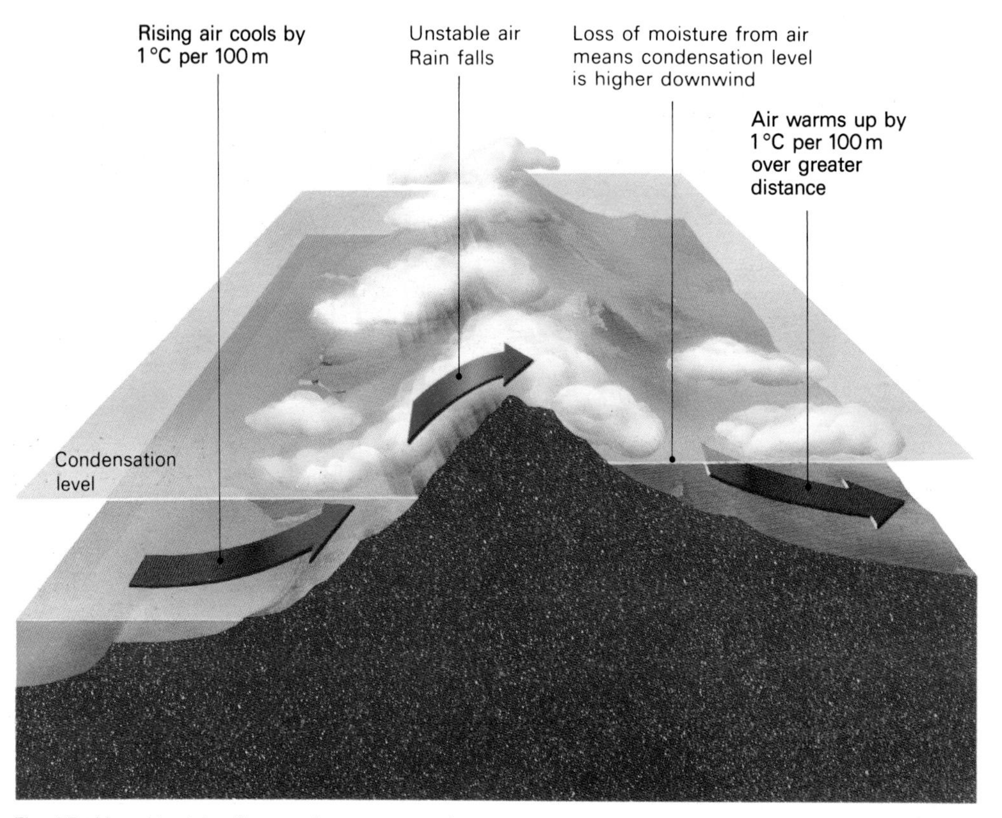

Fig. 4.7 Unstable air leading to rain over mountains

Fig. 4.6 Where the main air masses come from

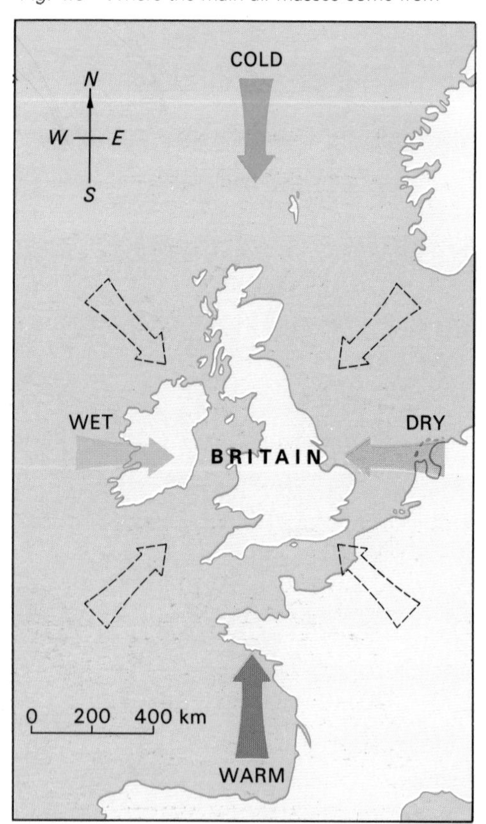

ii) A second way in which air rises and becomes unstable is through direct heating from the ground. Thunderstorms forming in the afternoon after a hot day are often caused in this way, but it is not a very common form of rain in Britain.

iii) Most of our rain comes from warm moist masses of air meeting cold air and being forced to rise over it. The meeting of these two masses of air is called a **front**.

Fig. 4.6 shows where these **air masses** come from and helps to explain why some are warm and moist while others are cold.

Air from the north comes from the polar regions and is thus liable to be cold. Air from the south has come from the tropics and tends to be warm. Air from the west has come from across the sea and will be humid and damp. Air from the east has come from the continent of Europe and will tend to be dry, for the North Sea is too narrow to have much effect on the dampness of the air.

7 Air from the south-west will be both warm and wet. Use Fig. 4.6 to help you say what air will be like from:
a) the north-west.
b) the north-east.
c) the south-east.

Because air from the south-west has come from the tropics and travelled across the sea, it is called tropical maritime. Likewise air from the north-east which has come from polar regions and across the continent of Europe is called polar continental.

8 What names do you suppose weather men give to air masses coming from:
a) the north-west?
b) the south-east?

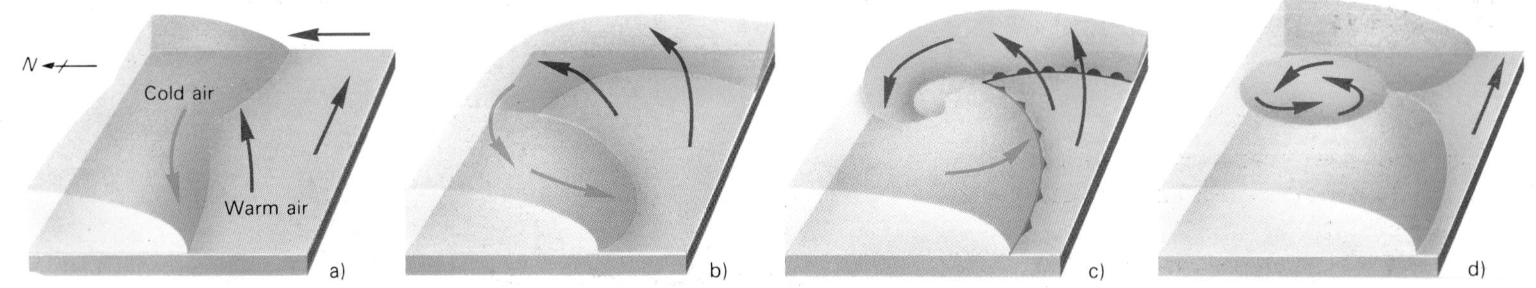

Fig. 4.8 How a depression is formed

Britain is thus a battleground for many air masses. The word 'front', describing a battle between two masses of air, was first used by weather men in Norway during the First World War. Where do you think they got the idea for the name?

In Fig. 4.8a two air masses are moving in opposite directions out in the middle of the Atlantic Ocean. The cold air mass is moving slowly westwards, while the warm air mass is moving slowly to the east.

One way of seeing what happens next is to put a pencil between your hands. If you move the right hand forwards and the left hand towards you, see what happens to the pencil. Air trapped between the two air masses begins to revolve in the same way as your pencil. It soon forms a swirling mass of revolving air as shown in Fig. 4.8b.

The air swirling round pulls outwards and lowers the pressure in the centre. We call this whirling mass of air a **depression**. A section of warm air begins to swing round towards the north and to climb above the cold air (Fig. 4.8c). This forms the **warm front** of the depression. The cold air swings round to the south and begins to edge in under the warm air. This is the **cold front**.

9 You can make a model of a depression. All you need is a sheet of A4 paper and to follow the instructions in Fig. 4.9.

Fig. 4.9 How to make a model of a depression

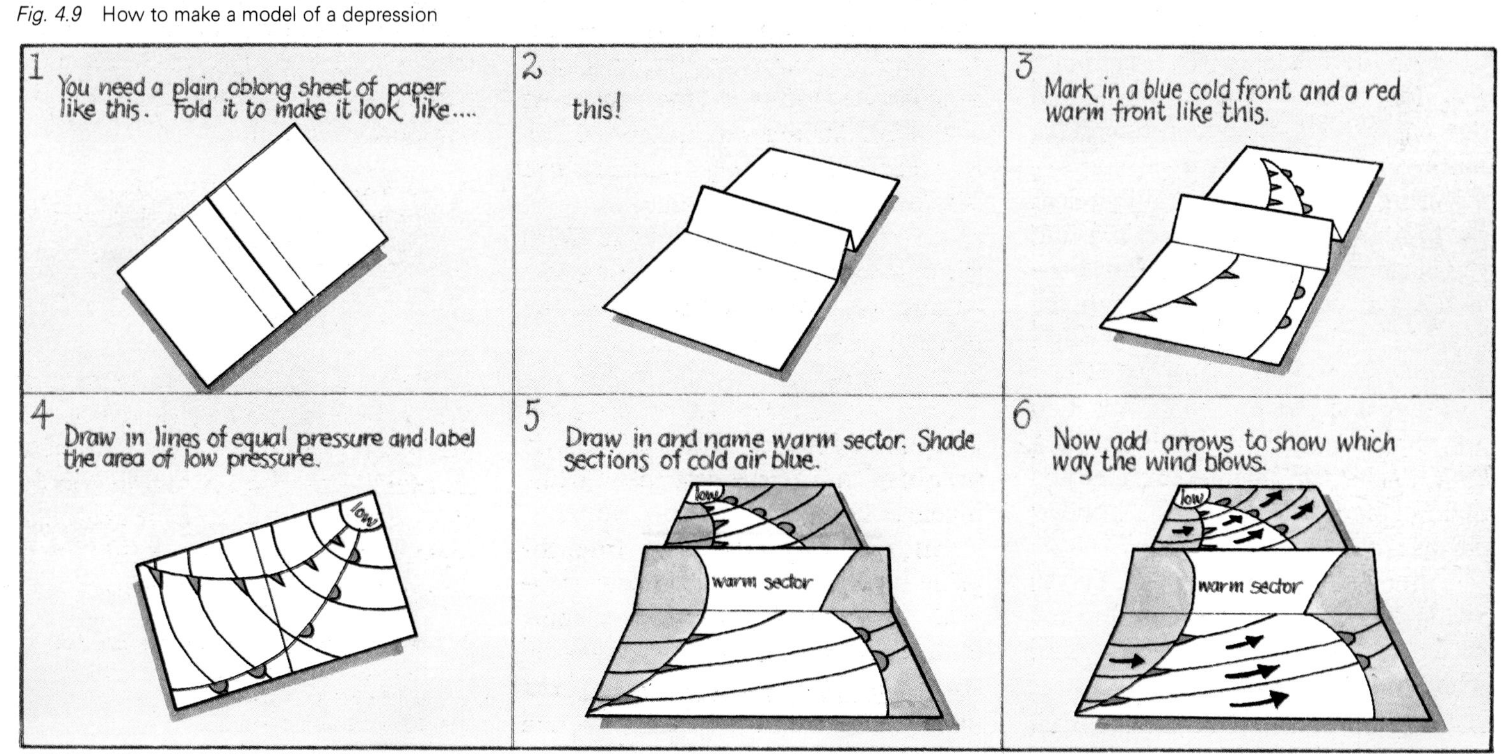

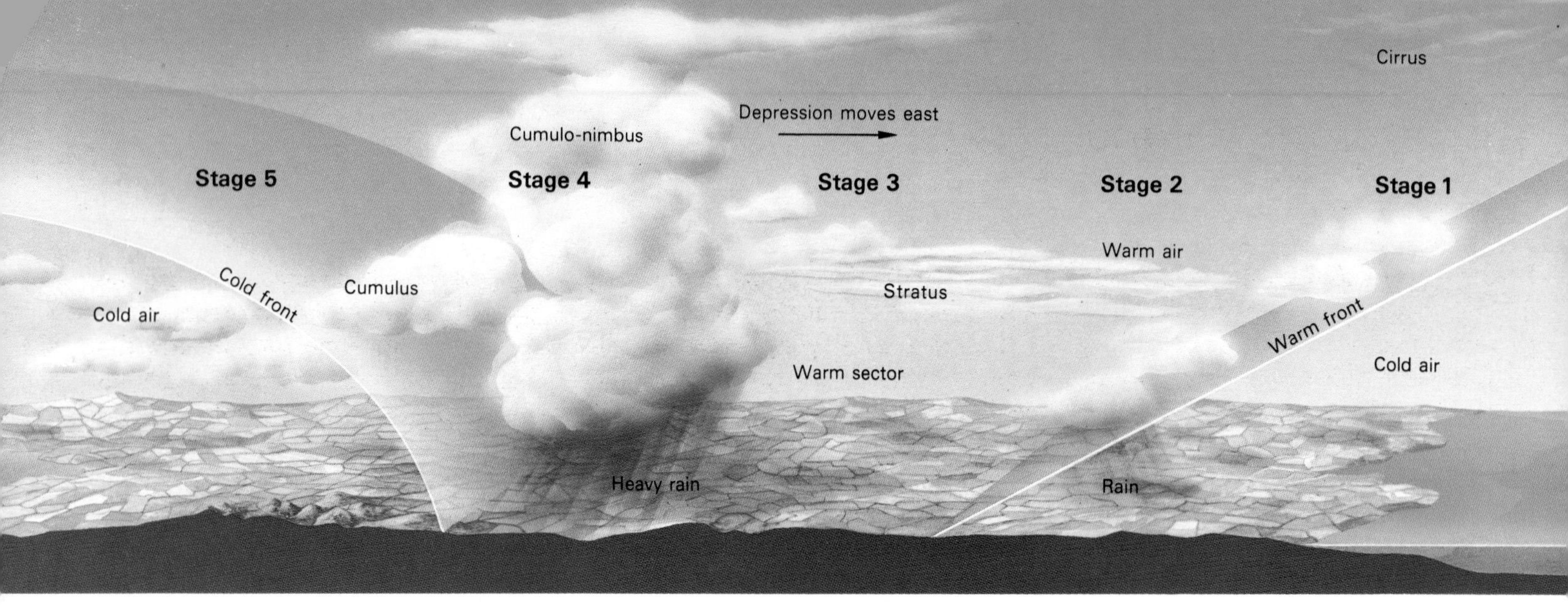

Fig. 4.10 What happens when a depression passes overhead

What exactly happens when a depression passes over Britain? The first sign of an approaching depression is high wisps of cloud as shown in stage 1 of Fig. 4.10. At the same time the sun will be shining, but the air pressure, measured by a barometer, will begin to fall.

The next stage is the passing of the warm front; drizzle or rain falls from a very dark cloudy sky as in stage 2 and the wind veers from south-west to south.

During stage 3 there may be clear sky in the warm sector or often a thin veil of cloud covers the sky. The shape of the sun can be seen through the cloud. The weather should be warm and muggy.

Stage 4 shows the cold front. Rain may be heavy and will often last for a long time. There may be thunder and hail falling from towering thunder-clouds which fill the sky.

After the cold front has moved away the air will feel fresher and cooler. The sky will be clear with white clouds like cotton wool as shown in stage 5.

Because the cold front moves across faster than the warm front, it will in the end catch up the warm front and lift the warm air right off the ground as in Fig. 4.8d. When this happens the result is an **occluded front** which normally gives drizzle and thick cloud.

10 a) Collect weather maps for a week from a daily newspaper (the *Times, Daily Telegraph,* and *Guardian* all have good examples).
b) Keep a weather diary for each day in that week, noting the weather at the time for which the map is drawn.
c) What weather do you find when the map shows:
i) a warm front?
ii) a cold front?
iii) an occluded front?
iv) an area of high pressure (an anticyclone)?

The amount of rain that falls is measured in a **rain gauge** (Fig. 4.11). If your school has a weather station it is worth inspecting the gauge. You can easily make your own to use at home with very simple materials like a funnel, a tin can and a jam jar.

A number of rain gauges spread over a town or a country area can show how rainfall varies from place to place.

Many people think that towns create their own rain. They say that the heat from the buildings warms the air and makes it rise, while the dust and soot provide more particles for water to condense round to form raindrops.

Fig. 4.11 Rain is caught in the copper can (12–cm diameter) and measured in the glass bottle

Water on the ground

What happens to the rain after it falls depends a great deal on what sort of ground it falls on.

If the rain falls on to the surface of a road or playground for example, most of the water will run across the surface; when it reaches a gutter, it will flow in a stream down to the nearest drain. Some water will stay on the surface as puddles to be slowly evaporated back into the air as water vapour.

11 You can use a simple method to measure the amount of evaporation.
a) Stick a ruler on end, using a piece of plasticine in a washing-up bowl. Fill the bowl with water to a depth of 60 mm.
b) Every day at the same time check the level of the water.
c) The difference will be the amount of water lost by evaporation.
d) After taking the reading, refill the bowl to the same level again.

If rain falls on to bare soil in a flower bed, the rough surface will stop the water from flowing away too fast. The soil will absorb the water like a sponge.

Some of the water will also evaporate from the surface of the soil, but soil which looks dry on top will often be quite damp underneath. The soil stores the water and this gives it time to sink further down into the rock below.

Heavy rain will compact the surface of the soil, however, and so more water will run off. Sometimes the rain will splash the soil, so that the soil gets washed away.

Grass on a lawn or sports field forms a thick mat over the surface. If rain falls on to a patch of grass, the soil is protected from the force of the rain and less soil will be washed away. The grass will intercept and store water, giving it more time to sink into the soil. The water will be protected from evaporating by the shade of the grass above.

12 Look at the three flow diagrams in Fig. 4.12. Decide which diagram shows what happens to water falling on:
a) soil.
b) grass.
c) concrete.

13 You can use a number of home-made rain gauges to work out how much water leaves and plants intercept.
a) Put one in the open, one under bushes or bracken, and one under trees.
b) Take readings over a period of a week or two to see how different they are.

Fig. 4.12 Flow diagrams for Exercise 12

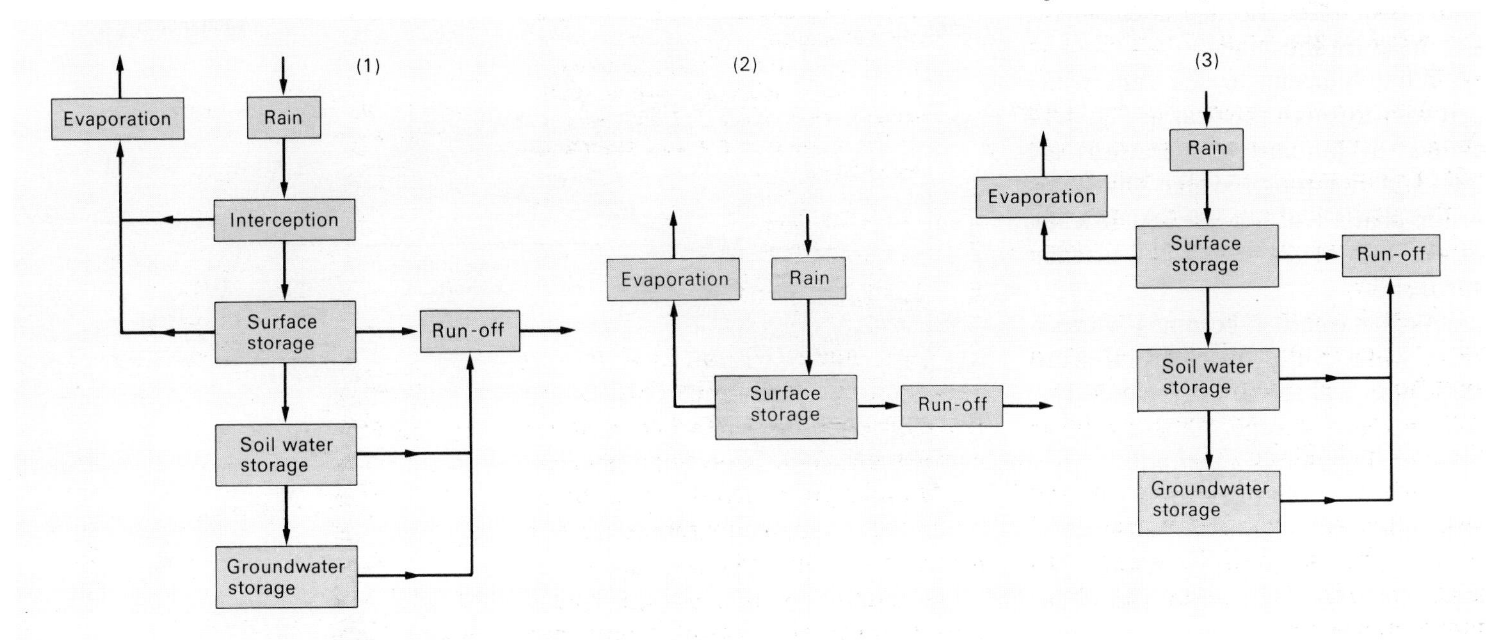

Fig. 4.13 shows what happens to the rain water or **precipitation** after it has fallen to the ground.

Some may be turned to water vapour and drawn back up into the air again. Some may be drawn up by plants and released through pores into the air. This is transpiration.

Some will sink or **infiltrate** into the soil as **soil moisture**; here it will be stored. Slowly, as the soil becomes soaked or saturated, some of the soil water will sink still further into the ground below to become **groundwater**.

Surplus water will either flow over the surface as **overland flow**, through the soil as **throughflow**, or through the rock below as **groundwater flow**.

Only a certain amount of water is available to be used by plants or to be evaporated from the surface. Often the air is dry enough for a lot more water to evaporate than there is about. In this case, if plants are to grow to their biggest, more water will have to be supplied by **irrigation**. The extra amount of water needed is called the **soil moisture shortage**.

What happens to the rain water can vary through the year as Fig. 4.14 shows. In January temperatures are low and not much water will evaporate; plants will not use much water at all. Rain or snow provides a water surplus.

As the weather becomes warmer, more water will evaporate and growing plants will start to use water from soil moisture storage. As the summer goes on, the stored water will be used up and water will have to be provided by irrigating the crops.

In the autumn there will be less loss of water by plants and fewer leaves to intercept the rain and stop it

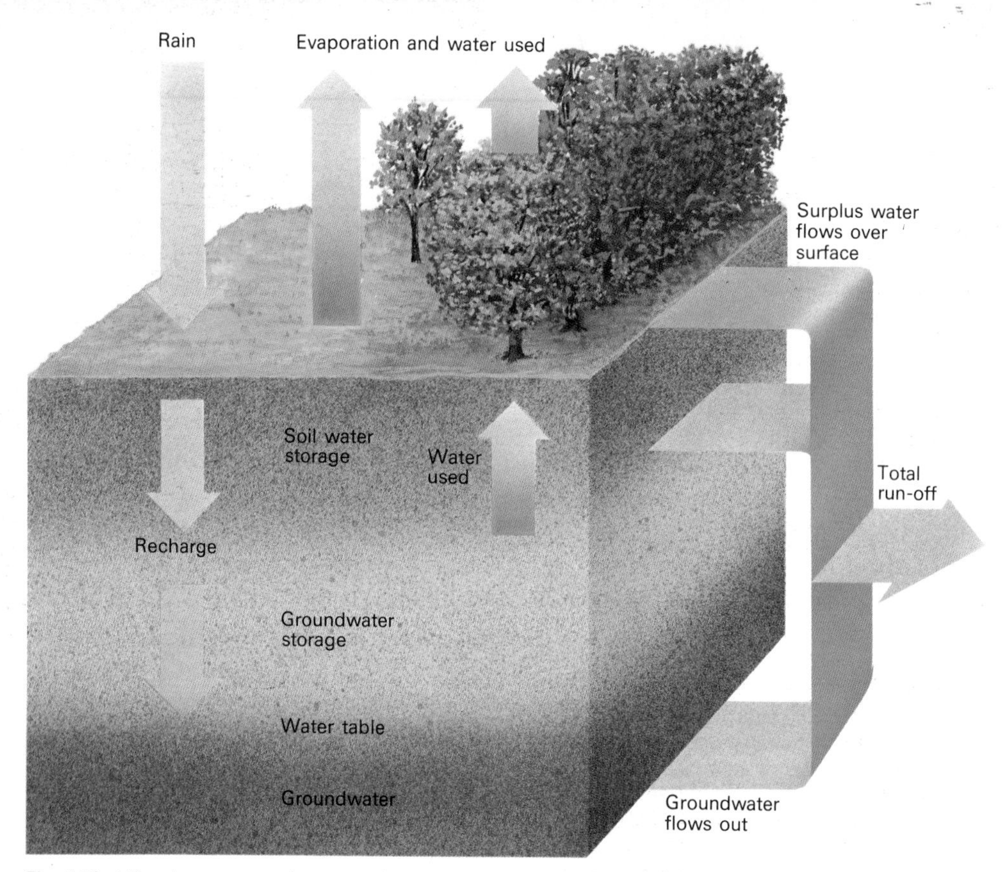

Fig. 4.13 What happens to rain water after it has fallen

Fig. 4.14 How water surplus and shortage vary through the year

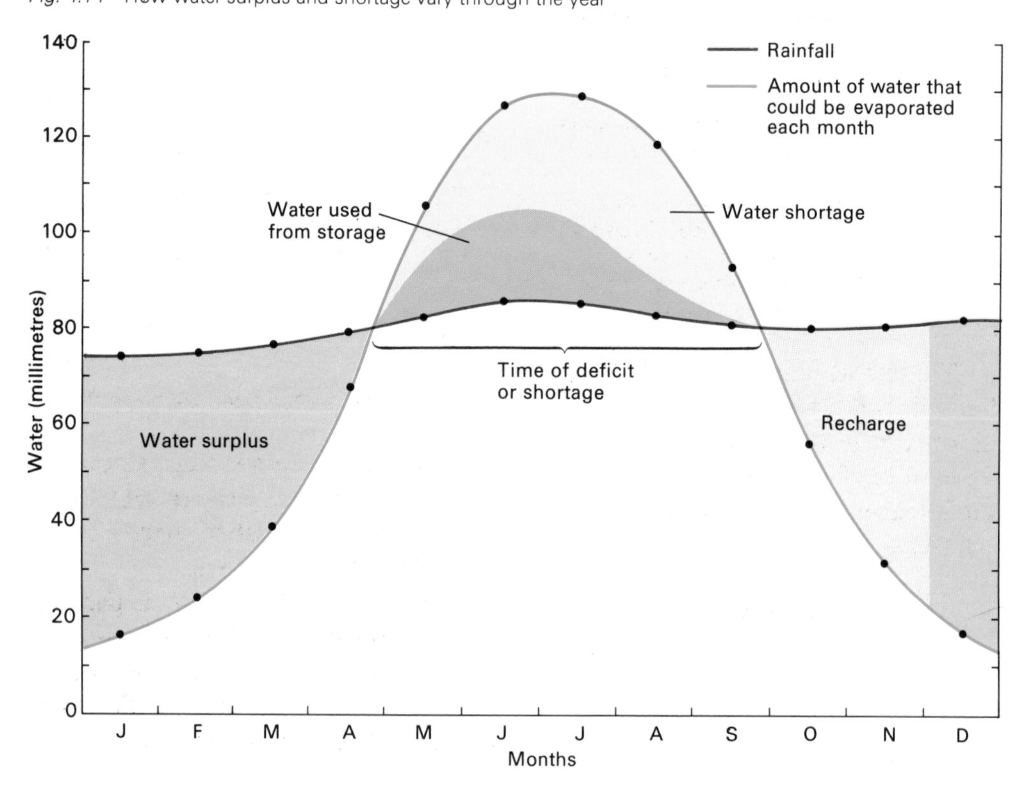

reaching the soil. Thus soil water stores can recharge. The soil will become soaked and saturated and there will be a surplus of water to flow over the surface again.

14 Use the figures in Fig. 4.14 to work out:
a) the amount of rain each month.
b) the amount of water lost from storage in the soil each month.
c) the amount of soil moisture shortage each month.
d) the amount of soil water surplus including recharge each month.

15 You can measure how much water sinks into the soil by using a hollow tube about 20 cm across and 30 cm deep. For example, a catering size tin with both ends removed or a piece of plastic drainpipe will make a suitable tube.
a) When you have pushed the tube half into the soil, stand a ruler inside it.
b) Pour water into the tube until it is, say, 50 mm deep. Use a stop watch to time how many millimetres the water sinks in a minute. It helps to do this a number of times so that you can check how accurate your readings are.
c) You can use your readings to draw a line on a graph like the red one in Fig. 4.16. You will need to change your rate per minute to a rate per hour by multiplying by 60.
d) On a rainy day you can use this graph together with hourly rain gauge readings, shown by the blue bars on Fig. 4.16, to work out the amount of water running off the surface.

The rate at which water sinks into the soil can vary from 20 mm per minute down to as little as 0·001 mm per minute. If each hour there is more rainfall than water sinking into the soil, the extra must run off the surface.

16 Using Fig. 4.16, how much water will run off after:
a) 2 hours?
b) 3 hours?
c) 4 hours?
d) 5 hours?

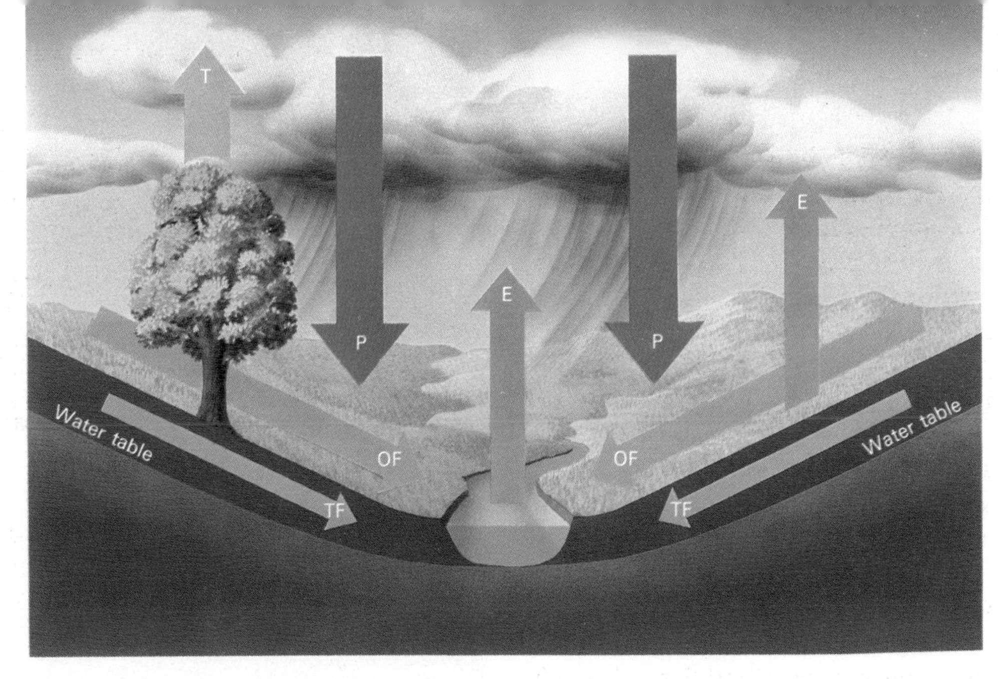

Fig. 4.15 Cross-section through a stream to show the flows and processes going on

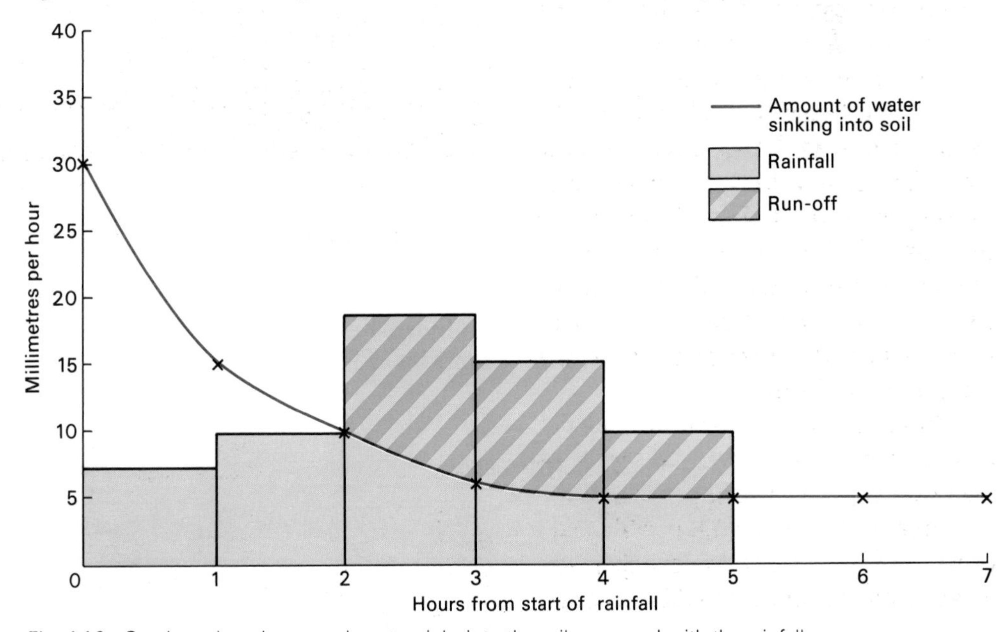

Fig. 4.16 Graph to show how much water sinks into the soil compared with the rainfall

The upper level of saturated ground is called the **water table** (Fig. 4.13). This rises when water seeps down from the soil above.

The water table tends to lie parallel to the surface and will thus be higher on the valley side than at the bottom of the valley. Water will only flow out of the soil into a stream where the water table is built up above the level of the stream. In winter this area gets larger until it covers about a third of the **catchment area** from which the stream obtains its water.

17 Using Fig. 4.15, write a sentence to describe what process each lettered arrow stands for.

Rivers

Fig. 4.18 The River Forth in Scotland: a winding or meandering river

For hundreds of years man has made rivers do work for him, using the river's energy to turn water-wheels. The energy used in a mill comes from the weight or mass of water and the height from which the water flows, the head of water.

Another source of energy depends on the speed at which the river flows. The greater the mass of water moving and the greater the speed at which it flows, the more energy there is to do work.

Rivers use their energy to do their own work. Nearly all the energy is lost through **friction** (95 per cent in some rivers is lost in this way).

The friction can be inside the river with swirls and eddies (**turbulent flow**) or it can be friction between the water and its bed and banks. The more rough and jagged the bed or banks of a river and the more weeds there are in the water, the greater the friction will be. Then there will be less energy to do any other work.

The shape of the channel cross-section will also have an effect on the amount of friction.

Fig. 4.17 Two river channels in cross-section

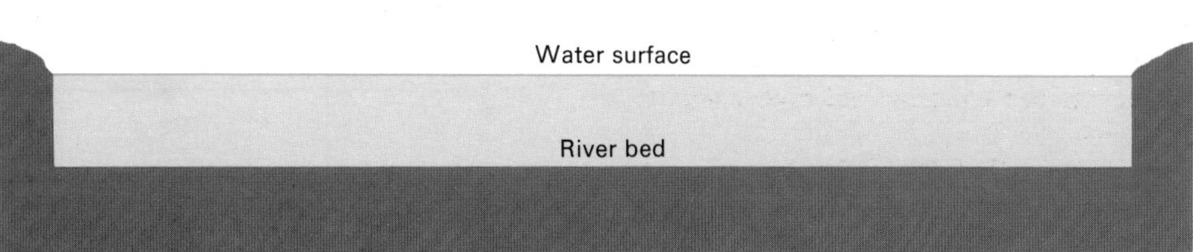

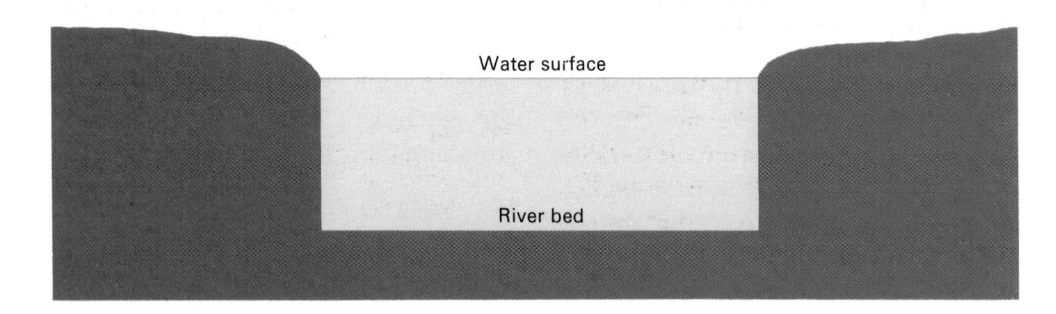

18 a) Which do you think will give more friction, the deep or the shallow channel in Fig. 4.17?
b) The two cross-sections have the same area: what is it?
c) What is the total width of bed and banks below the water of
i) the shallow channel?
ii) the deep channel?
d) Divide the width of the bed and banks into the area of the cross-section; the higher the number, the less energy is lost through friction.
e) Try out a few other channel shapes with the same area (square, semi-circle, etc.) to find out which gives the least friction.

A river uses its energy in three ways. First, it has to overcome the friction between its bed and banks. The energy the river has left over is used for two other types of work as shown in Fig. 4.19. The river has a **load** to carry and it also wears away or **erodes** its bed and banks.

A river transports its load in two main ways. Gravel, pebbles, or even

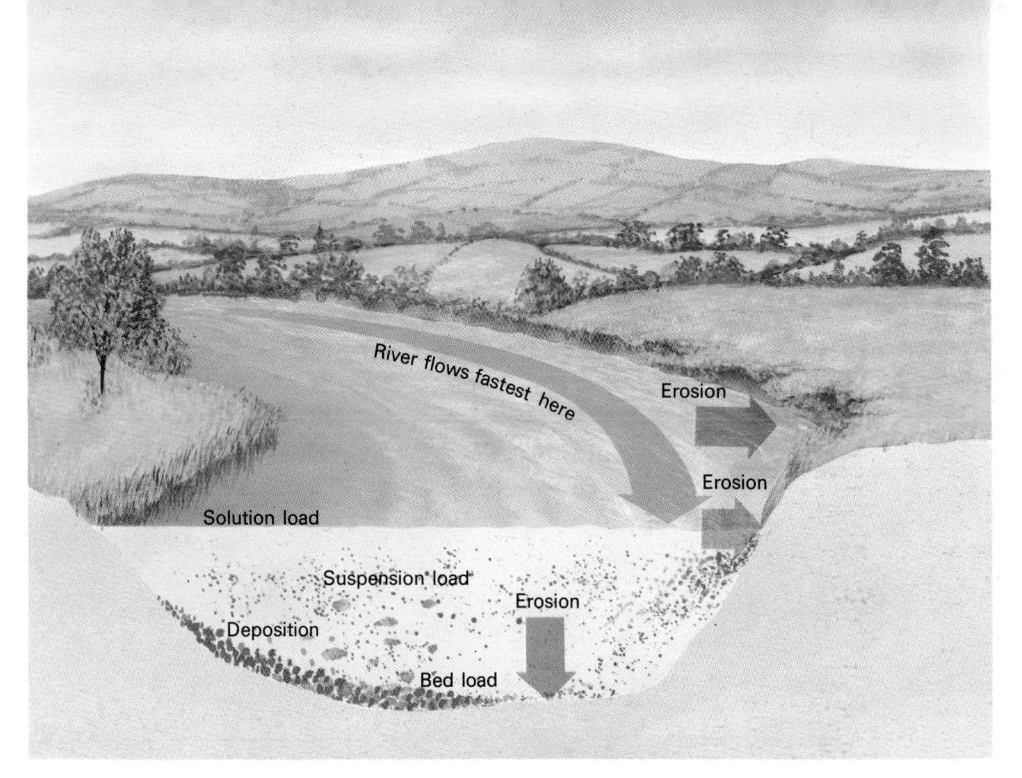

Fig. 4.19 A river does two kinds of work: erosion and carrying a load

boulders will trundle or bounce along the bed of the river as **bed load**. The greater the flow of the river, the greater the load it can carry.

Another part of the river's load floats along suspended as tiny particles of silt or sand. This is called the **suspension load**. One way of finding out how much material a river carries in suspension is to take samples in a special sample bottle as in Fig. 4.21.

19 We can measure the bed load of a river by sinking a box into the bed of the river with its top level with the river bed.
a) Weigh the amount of material that falls into the box over a few days.
b) Use sieves of varying sizes to see how much material of different size is carried as bed load.

20 Samples can be taken at various points down the course of the river or from the same point at different times to see whether the suspension load varies with the rate of flow.
a) After carefully weighing a piece of fine filter paper, use it to filter your sample through a funnel.
b) Let the sample dry and then weigh the filter paper and its contents.

The river in Fig. 4.18 is following a winding course. On the whole, it is only when people interfere with streams and rivers that they run in a straight line.

21 You can do an experiment to test this idea.
a) Obtain a flat tray and cover it with a thin layer of sand.
b) Put one end under a tap, tilt the board slightly and let the water from the tap trickle gently down the board.
c) Look carefully at the way in which the stream of water runs down the tray. Is it running down in a straight line or does it wind its way down?

The faster a river runs, the more energy it has to carry a load. You can demonstrate this by turning the tap on a little faster above the sand tray. Fig. 4.19 shows that when a river rounds a bend, the water flows fastest on the outside of the bend, scouring and eroding the outer bank of the river. Meanwhile, the slower running water on the inside of the bend may deposit its load of silt or gravel.

Fig. 4.20 A river eroding its bank

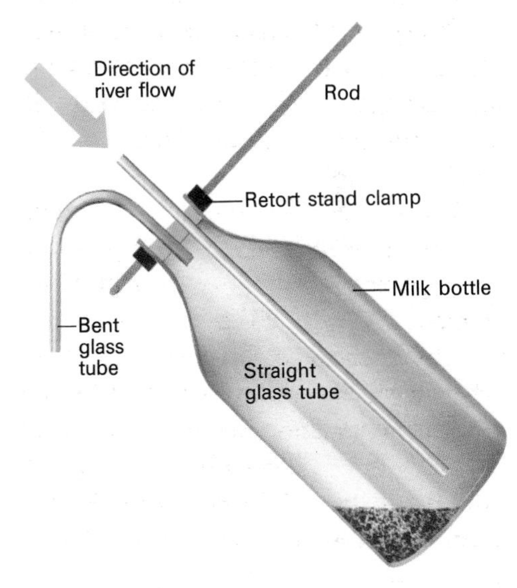

Fig. 4.21 A simply made sample bottle

22 a) Make a tracing or a copy of the river in Fig. 4.18.
b) Mark on your drawing where you would expect to find signs that the river was eroding its bank and where you would expect to find it depositing its load as sand or gravel.

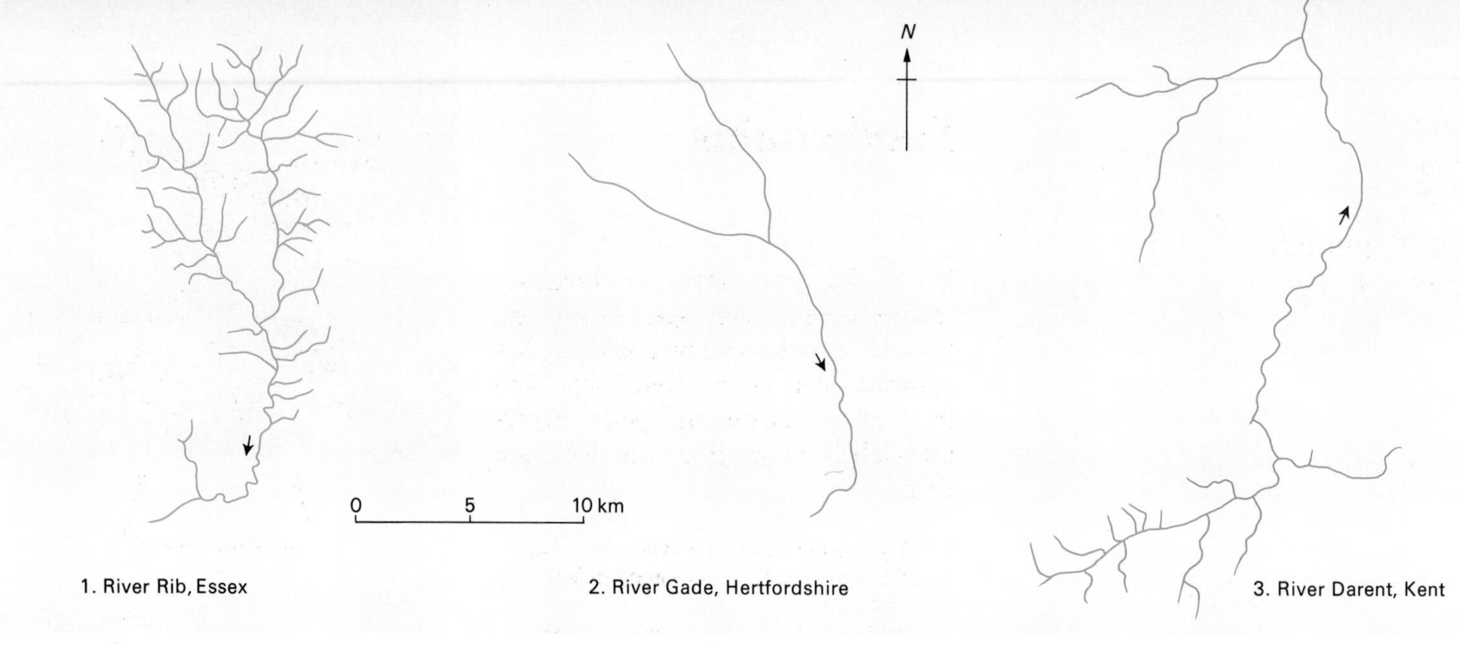

Fig. 4.22 Three rivers and their tributaries

Fig. 4.22 shows three different rivers and their tributaries. They form different patterns because they are flowing over different types of rock.

The first river is running over a rock that does not let water sink into it very easily: this type of rock is called impermeable rock. Very little rain water will sink into the soil or into the rock beneath. As a result there will be more overland flow and the river has a dense network of tributaries.

The second river is flowing over an area of rock which is very permeable so that there is not much water left to flow over the surface. This river has very few tributaries.

The third river is flowing through an area made up of bands of different types of rock. As a result the river pattern is like a trellis with the tributaries meeting each other almost at right-angles.

23 Which pattern in Fig. 4.22 is likely to form in areas of
a) chalk?
b) clay?
c) alternate layers of chalk and clay?

Weather station		Jan	Feb	Mar	Apr	May	Jun	Jul	Aug	Sep	Oct	Nov	Dec
A:	Temperature °C	4·1	4·2	5·6	8·1	10·8	14·1	16·3	16·5	14·8	11·3	7·7	5·3
	Rainfall mm	56	42	37	36	38	44	57	56	53	65	68	52
B:	Temperature °C	6·2	5·8	7·3	9·2	11·7	14·5	15·9	16·2	14·7	11·9	8·9	7·2
	Rainfall mm	105	77	73	55	65	58	71	80	82	94	115	115
C:	Temperature °C	4·3	4·4	5·7	7·0	9·3	11·6	13·3	13·3	11·8	9·3	6·9	5·5
	Rainfall mm	107	75	63	63	52	68	87	88	97	118	111	110
D:	Temperature °C	2·4	2·8	4·5	6·6	9·0	12·0	14·0	13·6	11·7	8·8	5·6	3·7
	Rainfall mm	77	54	52	50	62	53	92	73	65	90	91	78

Fig. 4.23 Table for Exercise 25

Summary exercises

24 Use the last chapter to help you explain the meanings of the following words:
radiosonde, barometer, unstable air, rain gauge, front, precipitation, depression

25 Fig. 4.23 gives the mean temperatures and rainfall for four different weather stations in south-east England, south-west England, north-west Scotland, and north-east Scotland.
a) Work out which figures are for which weather station.
b) Draw a bar graph to show the rainfall for each station.
c) Draw a line graph to show how the temperature varies at each station.
d) Why do you think the temperature and rainfall vary in these various parts of Britain?

5

Food, plants, and soil

Food chains

What did you have for breakfast today? Cereal? Porridge? Bread and toast? Egg and bacon perhaps? Fig. 5.1 shows a plate of egg and bacon and fried bread. Below the plate arrows and labels show how the food got there.

1 Try to work out what the arrows in Fig. 5.1 stand for. Here are some labels to help you:

made into bread	cured to make
fed to	fried to make
boiled to make	fed to
lays	

Copy the diagram and add a label to each arrow.

2 Draw a diagram like Fig. 5.1 to show the food chain for:
a) a bowl of cereal (with milk and sugar).
b) fish and chips.

As you can see people depend on plants to supply their food. These are either plants for them to eat or to feed the animals they eat. This link is called a **food chain.**

Fig. 5.2 shows that you need a large mass of plants to feed the small plant-eating animals, and a large mass of small plant-eating animals to feed one small meat-eating animal. In turn you need a mass of small animals to feed one large meat-eater.

So it is clear that a very large amount of plants are needed to keep one large meat-eater alive. Most food chains are quite short with about three levels: plant-eaters, small meat-eaters, and a large meat-eater.

Animals that eat meat are called **carnivores** and those that eat plants are called **herbivores**. Some animals eat both plants and other animals; they are called **omnivores.**

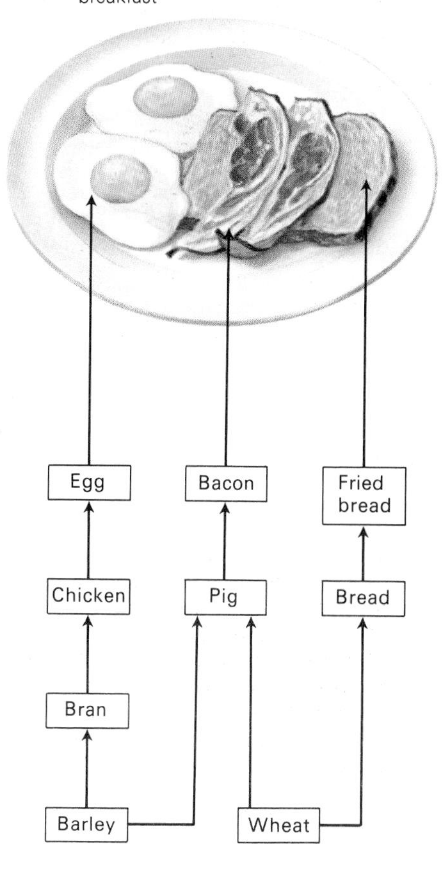

Fig. 5.1 The food chain that leads to someone's breakfast

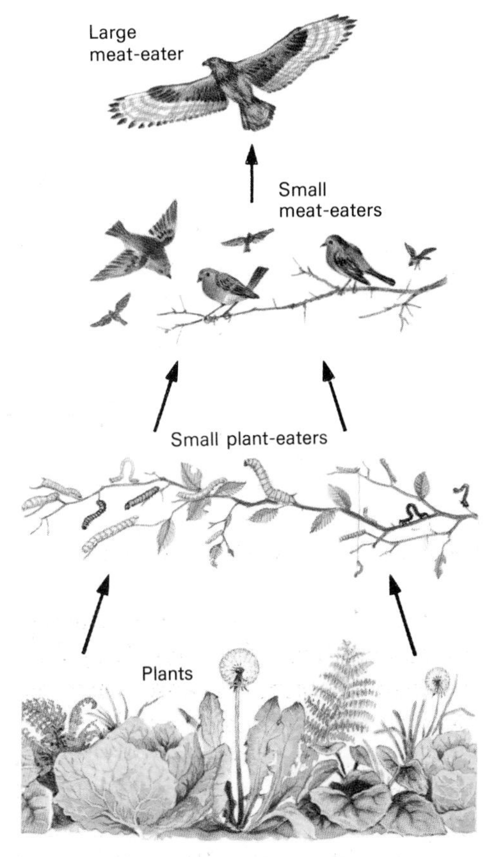

Fig. 5.2 A food chain

3 Make lists under three headings of all the pet animals you can think of to show whether they are herbivores, carnivores, or omnivores.

We all depend on plants for all our food. Things that happen to plants at the bottom of the chain can affect animals at the top of the chain.

For example, when poisons such as DDT are used, the DDT not used up in killing insects is washed into the soil or into ponds and streams. Plants draw up the water and a small amount of DDT stays in their leaves and stems.

Animals eating lots of these plants absorb DDT. Animals higher up the

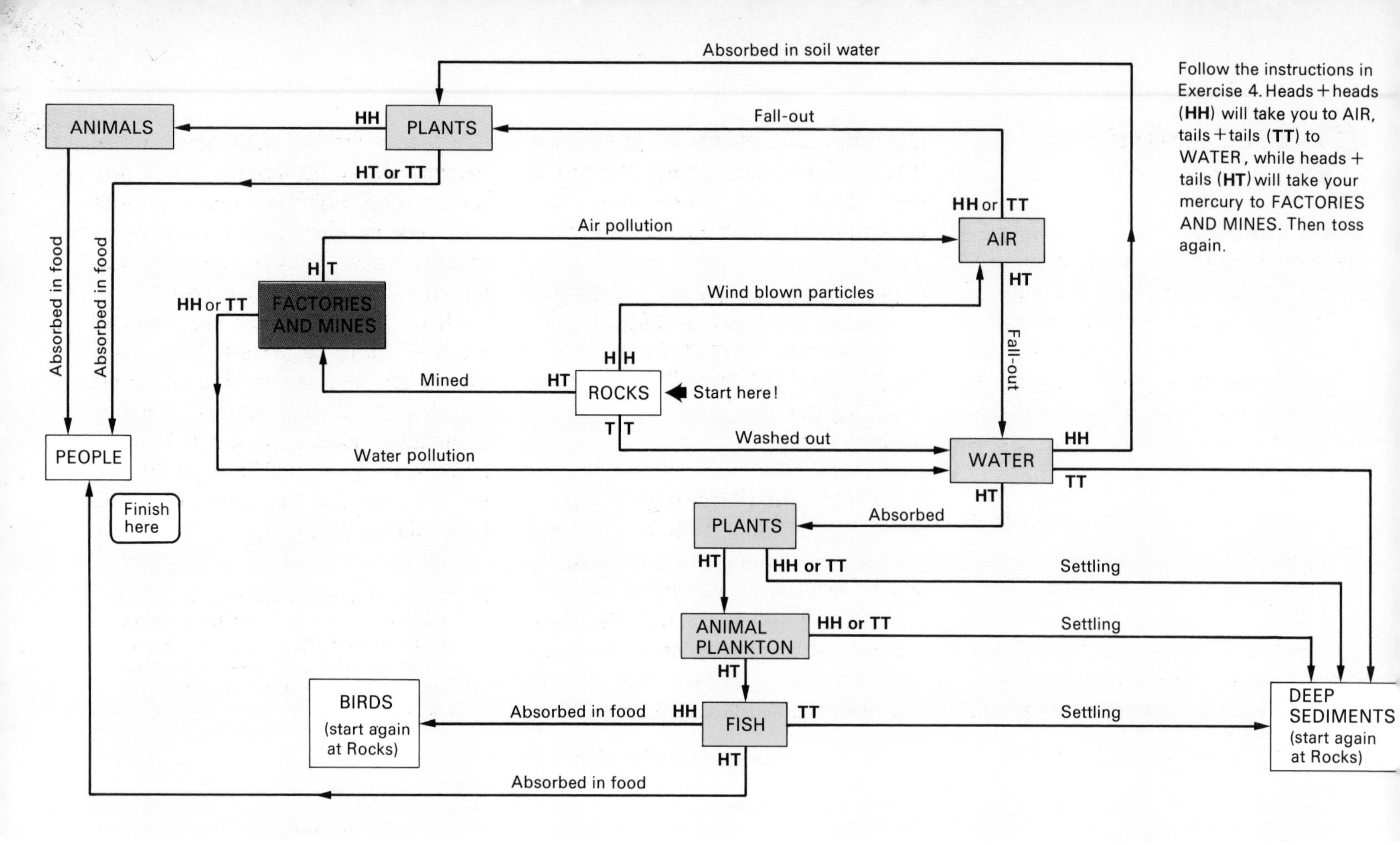

Fig. 5.3 The mercury maze game

food chain which eat animals with DDT in their flesh will also receive DDT. This can reach very high levels.

As a result the animals or birds at the top of the food chain may die or suffer ill-effects. For example, in birds this can lead to very thin eggshells that break before they hatch. Rare birds like grebes, pelicans, and peregrine falcons will have less chance of hatching young chicks and so their numbers fall.

Still more serious to people are the effects of mercury as a poison. It is used to make chlorine, and in farming to kill fungi. Fig. 5.3 shows the many different ways in which mercury can reach people.

4 Fig. 5.3 is in the form of a board game. You can play in small groups, using two counters for each person. The counters represent drops of mercury. You also need two coins to toss to decide which paths your drops of mercury will follow.
a) Put your counters in the box labelled 'rocks' and toss the two coins to decide which box each mercury drop will move to in turn.
b) Then toss the coins again to decide which box each drop moves to next. As each drop moves along a path, make a copy of that part of the diagram in your exercise book.
c) You have finished the exercise when your mercury has reached 'people'. The winner will be the person with the most complete diagram.
d) Draw in any other paths to make your diagram complete.
e) Can you think of any ways to stop the mercury reaching people?

Mercury poisoning is sometimes called Minamata disease. This is because 52 people died from mercury poisoning and over 900 fell very ill in a Japanese fishing village on Minamata Bay in 1953.

A giant chemical factory had been emptying waste mercury into the bay and fish caught by the Minamata fishermen were highly poisonous. This was especially true of shell fish such as oysters. They had over 70 000 times the amount of mercury in them as the water in Minamata Bay.

Waste from factories and from farming has caused severe mercury poisoning in Japan, Sweden, and the United States.

Food for plants

The food webs in Fig. 5.6 and Fig. 5.8 show that plants were at the root of early man's food chain. After many thousands of years of invention and change in farming, plants are still at the base of our food chain today. Fig. 5.4 shows how all our food comes from plants.

☆5 a) Describe what Fig. 5.4. shows in your own words.
b) What are the main differences between the diagrams in Fig. 5.4 and Fig. 5.6?

The three main chemical foods which plants need are nitrogen, phosphates, and potassium. Nitrogen (N) helps the plants to grow their leaves. Plants need phosphates (P) to help them grow strong roots and potassium (K) to help them produce fruit.

6 Farmers are very careful to apply the right amounts of nitrogen, phosphates, and potassium to their crops. Which chemicals would you expect them to apply most to:
a) pasture?
b) potatoes?
c) tomatoes?

Where does the nitrogen used by plants come from? In the case of crop land much comes from the farmer adding fertilizer or manure to the soil.

When plants or animals die, they begin to rot or **decompose** as shown in Fig. 5.5. The bacteria which cause them to rot produce ammonia. This is one reason for the nasty smell. Other bacteria thrive on ammonia and produce nitrites which are highly poisonous. Yet another group of bacteria change these nitrites into nitrates which are a useful plant food.

If too many nitrates are provided for the plants to absorb, they wash into lakes and rivers. Water plants called algae then grow very rapidly. Sometimes lakes turn green with so much plant growth. The algae cut out the light and use up the oxygen in the lake, killing all the fish.

7 Making fertilizers which provide nitrogen uses a lot of energy. Therefore wasting nitrogen is a waste of valuable energy resources. Produce a leaflet explaining to farmers why adding too much nitrogen could be harmful.

Fig. 5.4 How plants are the basis of all farming

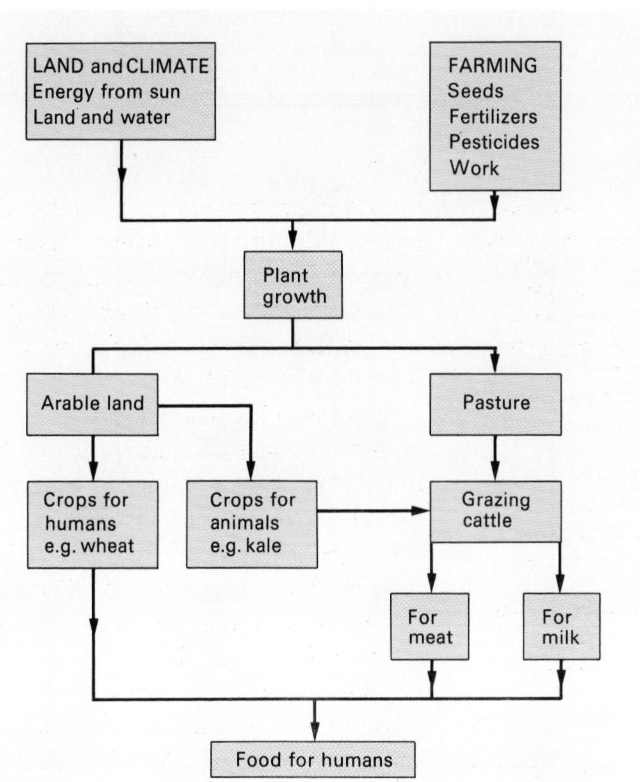

Fig. 5.5 The nitrogen cycle

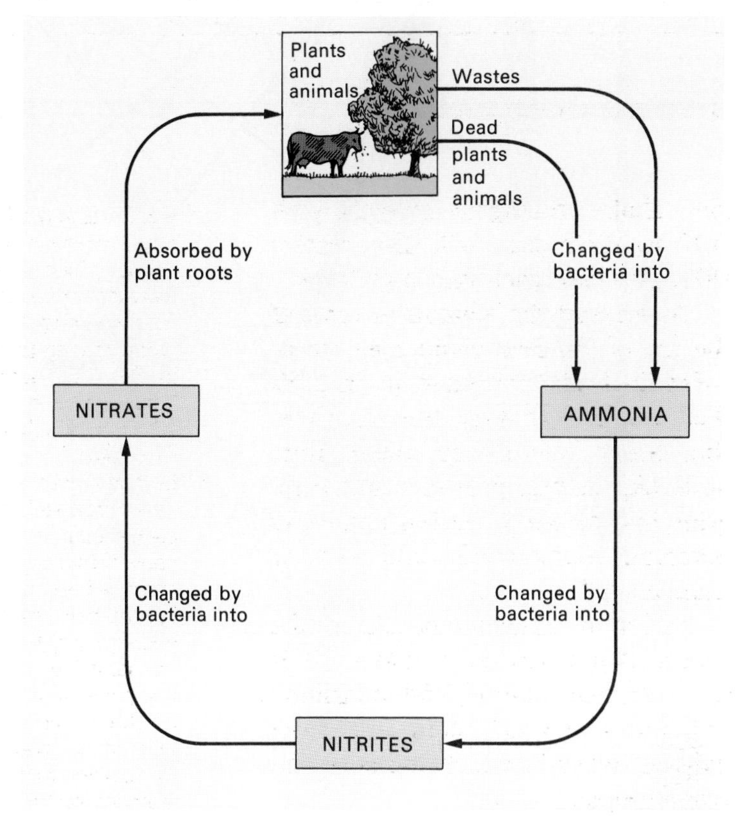

Food webs

By studying and counting the bones found in ape-man's camp sites in East Africa, we can build up a picture of what he ate over a million years ago. We do not know what plants he ate because there are no remains, but there are the fossil half-eaten bones of antelopes, zebras, baboons, sabre-toothed tigers, and all the other animal life shown in Fig. 5.6.

The bones also suggest that it was not only man who ate other animals; there is more than one food chain shown in Fig. 5.6. It shows a **food web**—lots of food chains linked together.

Fig. 5.6 Ape-man's food web

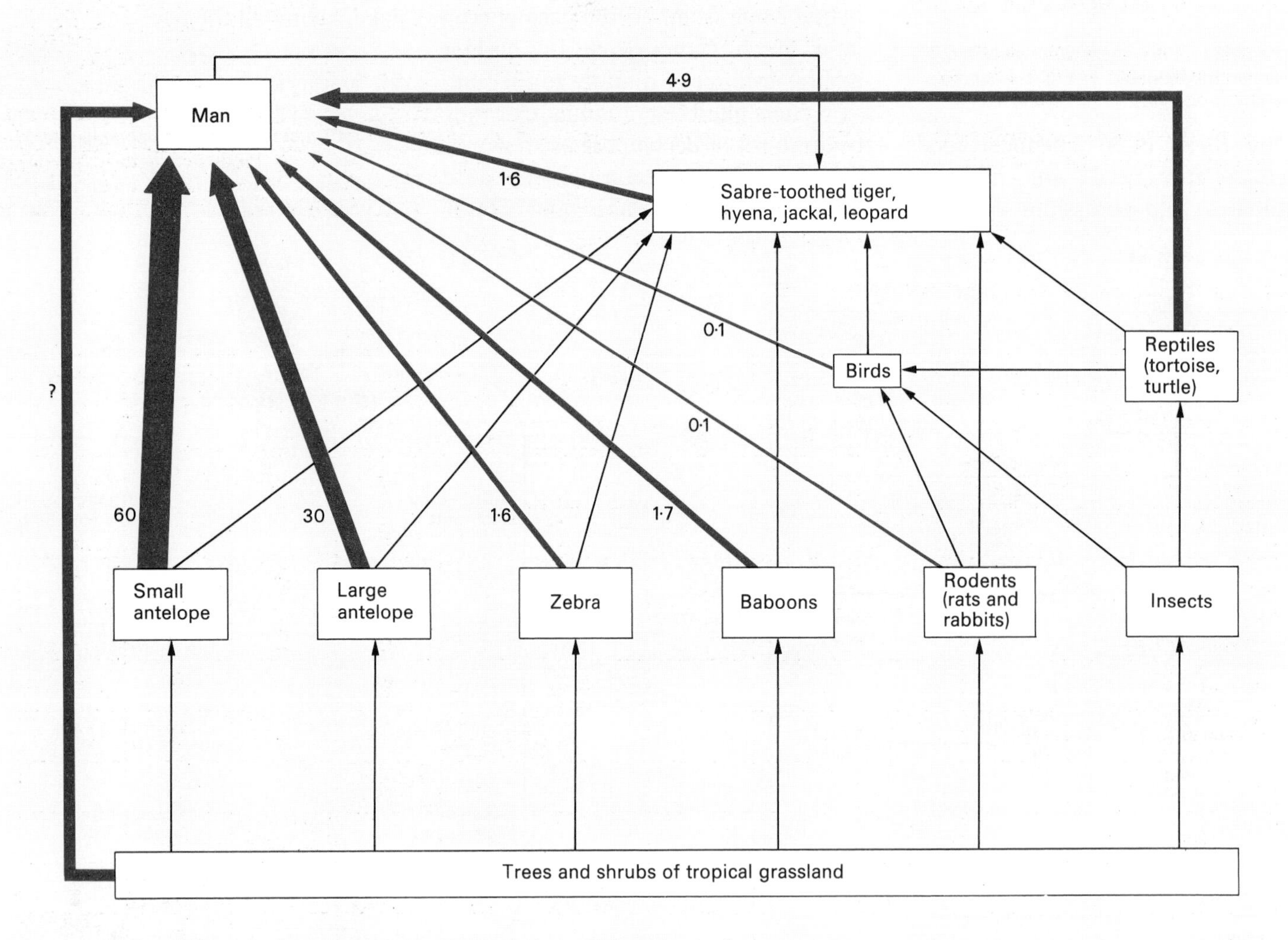

The numbers show how many bones of different kinds of animal were found in every hundred on sites in Africa

is called a **podsol**. The diagram shows what this soil looks like if we dig a pit or a trench and look at the sides. Many clear layers are formed in the soil. They can tell us a great deal about what has been happening in the soil.

The top layer is made up of pine needles, below which a thin layer of black humus can be seen. Under this a grey sandy layer shows where the minerals have been washed away down the soil to be left as a rust-coloured layer rich in iron. Below is a yellow layer of lighter minerals.

Because there is so little plant material in a podsol, there are few animals living in the soil to mix up the layers. There is not much nitrogen, so the soil is not very fertile.

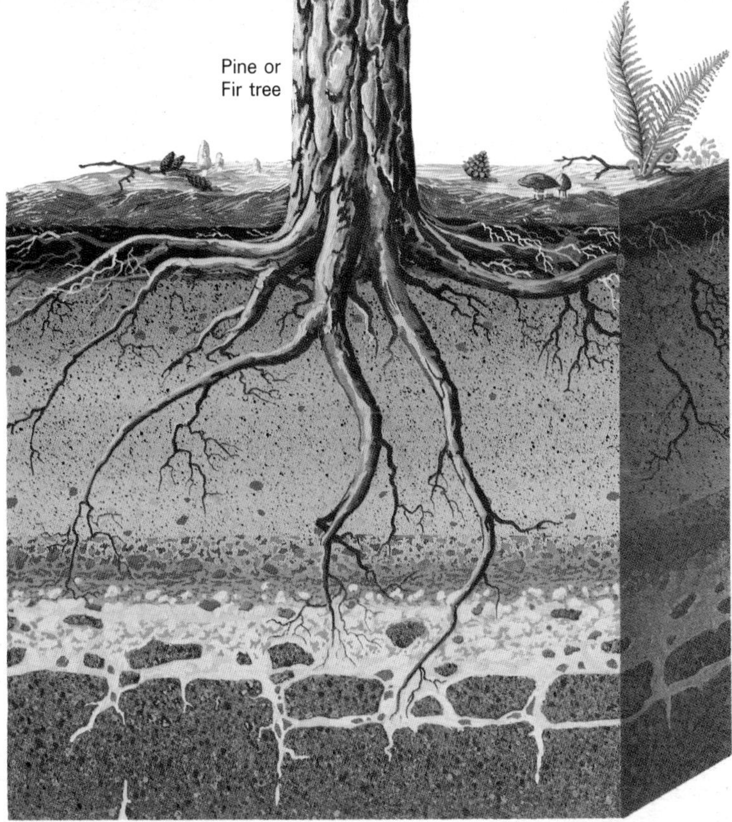

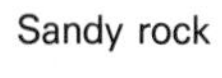

Fig. 5.10 A section through podsol soil

Soil takes hundreds of years to form; we can often see it starting on roofs or on paths which are not walked on very often. If the air is damp, lichens will grow on rocks, concrete or tarmac and particles of dust will be trapped in the lichens' spongy growth. After a while moss will be able to grow, forming thicker spongy masses to store water and dust particles.

The lichens and moss will have arrived as spores blown by the wind. Other seeds will soon arrive too so that we sometimes find plants growing on buildings. Grass, rose-bay willow herb, buddleia, and even silver birch trees may do this.

Once grass has arrived, it will begin to form turf so still more dust and moisture can be trapped. Earthworms and other animals may move in and gradually the turf will become thicker. The earthworms will make the soil deeper by bringing their worm-casts to the surface.

In this way whole ruins of buildings can be buried under the soil. This is why archaeologists so often have to dig to make their finds.

Fig. 5.11 Valerian growing on a ruined barn

12 What examples of soil forming and plants beginning to grow can you find in your area? Look especially in places where buildings have been demolished and on old walls, tennis courts, old railway lines, roofs, and gutters.

Plant succession

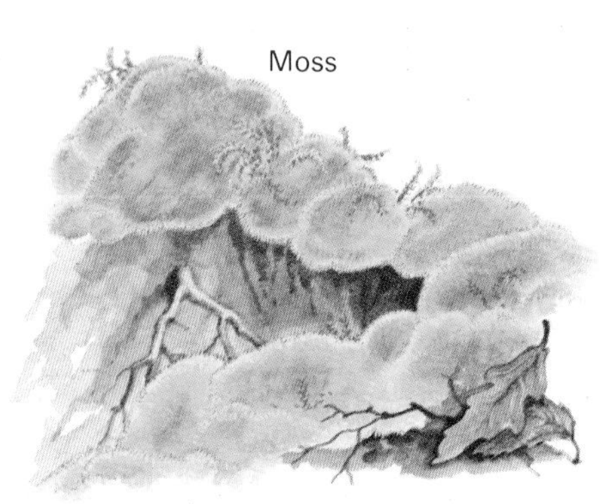

Fig. 5.12 Marram grass growing on the dunes at Studland Bay

Grass grows taller than moss and lichens. The shorter plants will not be able to survive because they will not receive enough light. The grass is said to have succeeded the moss; this process is known as **plant succession.**

We can see the beginnings of this in a badly weeded flower bed or an allotment. Often moss and lichen will begin to coat the soil and then quick-growing weeds like dandelions, chickweed, or groundsel as well as grass take over.

If the plot is not looked after for a few years shrubs and bushes such as buddleia will grow and even small trees such as sycamore. All these plants are shown in Fig. 5.13.

Sand dunes are a good example of a bare surface being colonized by plants. When the wind blows across a wide sandy shore, it can pick up grains of sand and build them into heaps of sand or **dunes** as shown in Fig. 5.12.

Sand from the sea at Studland Bay in Dorset is carried up the beach by waves during winter gales when the wind is in the east. When the sand

Fig. 5.13 Some plants that invade a flower bed

Dandelion

Groundsel

Chickweed

Grasses

Sycamore

Buddleia

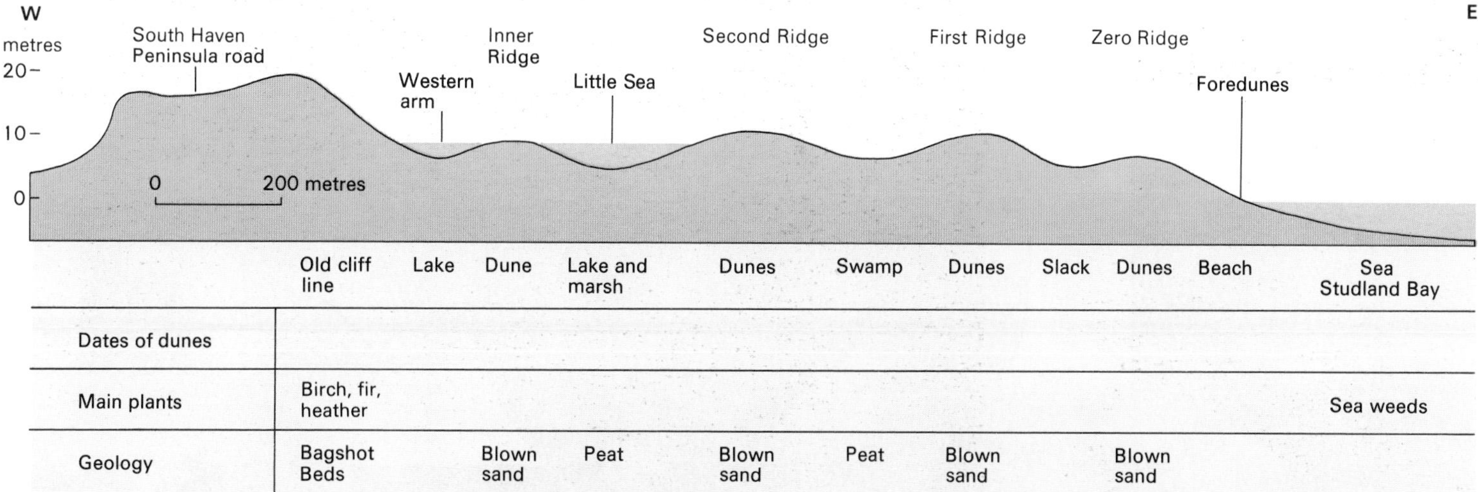

Fig. 5.14 Cross-section of South Haven Peninsula showing the dune ridges

dries out after the storms, it is blown inland and will form dune ridges. If you walk across the dunes at Studland Bay, you cross a number of ridges westward from the sea (Fig. 5.14).

By looking at old maps, like those in Fig. 5.15, we can see when these ridges were formed. In the year 1600 there was only a thin line of land called South Haven **Peninsula** jutting out into the sea. It was made up of pebbles cemented together called the Bagshot Beds. By 1721 sand dunes had begun to build up on the eastern side of the peninsula to form what is now the Inner Ridge of the Little Sea.

The early Ordnance Survey map drawn in 1849 shows two more ridges. They were given the names First Ridge and Second Ridge in 1933 because they were then the first and second ridges people came across when they set out from the seashore to walk westwards.

By 1947, however, another ridge had formed further east and so a name was invented for this: Zero Ridge. Just to the east of Zero Ridge, smaller dunes have begun to appear; these are called the foredunes.

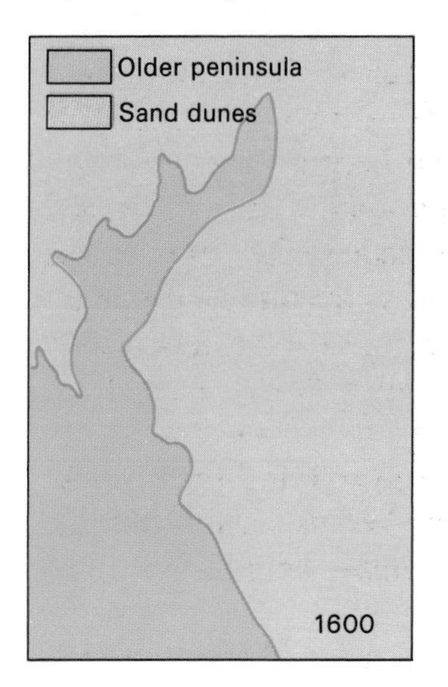

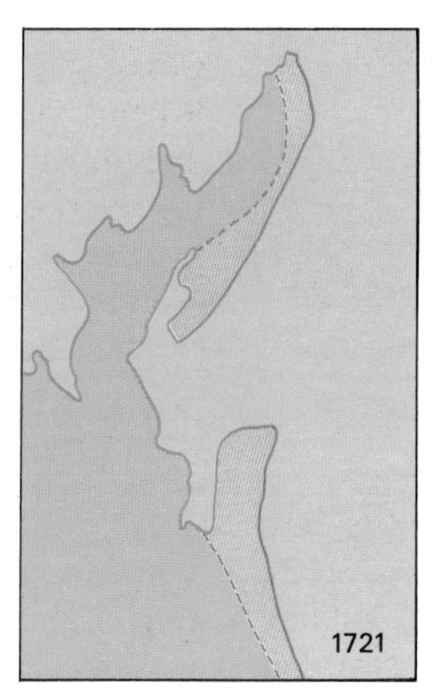

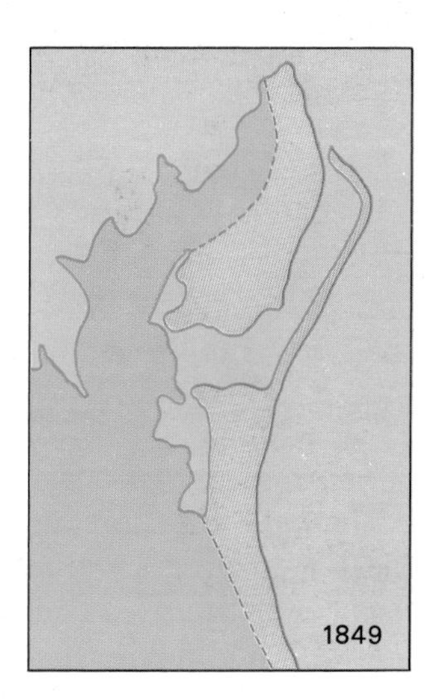

Fig. 5.15 Maps to show the development of South Haven Peninsula

13 If the foredunes were to grow into yet another dune ridge east of Zero Ridge, what name would you think should be given to it?

14 On a copy of Fig. 5.14 fill in the dates for the formation of each of the dune ridges.

The foredunes just above the beach at Studland Bay have sea couch growing on them. This helps to bind the sand grains together with its roots. Sea couch can stand being covered by the sea at very high tide.

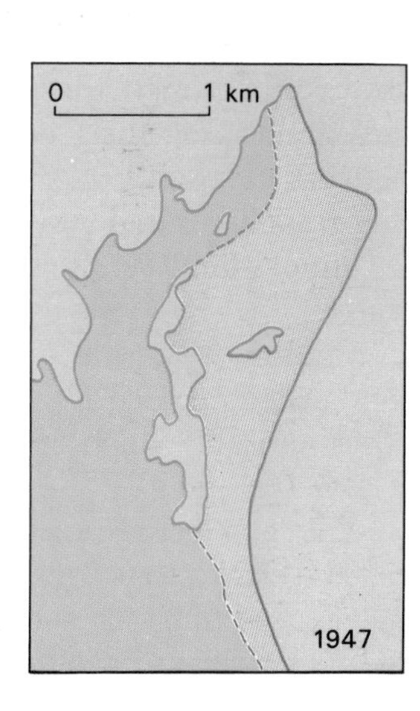

a) Sand sedge
b) Red fescue with pink thrift behind
c) *(right)* Creeping willow
d) Alder
e) *(above)* Sallow
f) Birch

Fig. 5.16 Plants that grow on sand dunes

Behind the foredunes on Zero Ridge marram grass grows (Fig. 5.12). It is better than sea couch at binding the sand and fixing the dunes so that they do not blow away. If the marram grass is covered in sand, the grass grows up through the sand, binding the grains together with a dense network of strands of grass. Although this grass can stand very dry weather, it cannot stand being covered in salt water.

On the west side of Zero Ridge, sand sedge and red fescue are found, other grasses which bind the sand (Fig. 5.16a and b).

The gap between dune ridges is called a **slack**. Zero Slack, west of Zero Ridge, used to be the shoreline during the 1930s. Now creeping willow (Fig. 5.16c) trails across the damp ground with trees of alder, sallow, and birch (Fig. 5.16d, e, and f).

First Ridge is covered in bell heather (Fig. 5.16g) with moss and lichen in between. The only sand to be seen on the surface is on the pathways where people have trampled the heather and exposed the sand.

In between First Ridge and Second Ridge lies East End Marsh Tongue where cotton grass is the main plant to be seen, a sure sign of swampy conditions (Fig. 5.16h).

Second Ridge has heather and gorse. Marram grass has been planted to patch up the areas where tourists have worn away the heather to expose the sand. A thin black line of humus can be seen on the edges of these sandy areas. This is the first stage of soil forming.

The western shore of the Little Sea has birch trees growing with gorse scrubland. This is the sand dune formed by 1721. Beyond this is the western arm of the Little Sea. Its western shore is the old line of cliffs which was the coastline in 1600 (Fig. 5.15).

15 On your copy of Fig. 5.14 write in the names of the main plants growing on each of the dune ridges.

☆**16** Which plants would you have expected to find growing on the oldest dune ridge:
a) in 1721?
b) in 1947?
c) today?
Draw a diagram like Fig. 5.13 to show the plant succession on the sand dune through the years.

g) *(top)* Bell heather with yellow gorse

h) *(above)* Cotton grass

Plant succession on Krakatoa

Another example of plant succession is to be seen on Krakatoa (Chapter 1, p. 19). After fifty years the 30 metres depth of ash left by the eruption had been covered by rain forest similar to that on the island before the blast (Fig. 5.17).

17 Use Fig. 5.17 to write a description of what explorers might have seen on their visits to the island in:
a) 1883.
b) 1886.
c) 1908.
d) 1934.

☆**18** Use the following description to help you produce a table like Fig. 5.17 to show plant succession on the island of Surtsey, off the coast of Iceland.

'It may be assumed that the Surtsey moss will develop a thick continuous carpet. This moss carpet will then collect dust and minerals, and nutrients will be deposited in the dead mat of moss which will form a layer of humus in the juvenile soil.

'The accumulation of humus will then cause moisture to be retained on the lava surface. The pioneer lichens will occupy the higher ridges of the lava and gradually corrode its surface, providing better anchorage for other plants.

'Gradually a heath vegetation with sedge, crowberry, and low-growing willows may invade the moss carpet in the most sheltered areas. But it is highly unlikely that the island will ever obtain a climax vegetation of birch as the lava flows of the mainland, because of the frequent salt spray and the heavy storms.'

(S. Fridriksson: *Surtsey: Evolution of Life on a Volcanic Island*, Butterworth)

Fig. 5.17 Plant succession on Krakatoa

Year	Total number of plant species	Coast	Lower slopes	Upper slopes
1883	0	Volcanic explosion; all life killed; hill slopes deeply gullied by rain		
1884	0		No life	
1886	26	9 species of flowering plants	Ferns and scattered flowers; blue-green algae on ash	
1897	64	Woodland	Dense grasses	Dense grasses and shrubs
1908	115	Wider belt of woodland with more species, shrubs and coconut palms	Dense grasses up to 3 metres high, woodland in ravines	
1919	214	(Similar to 1908)	Scattered trees in grassland with thickets in ravines	
1934	271	Woodland	Mixed woodland largely replaces grass	Woodland with smaller trees

Summary exercises

19 Use the last chapter to help you explain the meanings of the following words:

food chain	food web
carnivore	nitrogen
herbivore	humus
DDT	leaching
Minamata disease	podsol

20 Put the following list of plants in the order in which they are likely to colonize sand dunes: marram grass, birch, creeping willow, sea couch.

6

Erosion

Weathering

The Devil's Marbles in Australia (Fig. 6.1) are rocks made of granite similar to those in Chapter 2 (p. 23). They form a tor, a pile of rounded granite boulders.

These rocks were formed deep in the ground (Fig. 6.2a). It took millions of years for the rocks above the granite to be washed or blown away. As the pressure from above was gradually reduced, cracks or **joints** began to appear in the granite (Fig. 6.2b).

Once the surface was worn down to reach it, the granite slowly rotted or **weathered** along the joints (Fig. 6.2c). Because the corners of the blocks were exposed on three sides, the corners crumbled more rapidly, rather like sugar cubes dissolving in a cup of tea. The square blocks of rock weathered into rounded boulders. Finally, the weathered rock was blown or washed away to expose the granite boulders (Fig. 6.2d).

Fig. 6.1 Weathering on the Devil's Marbles; they are 400 km north of Alice Springs in central Australia

a) Overlying rocks

Granite

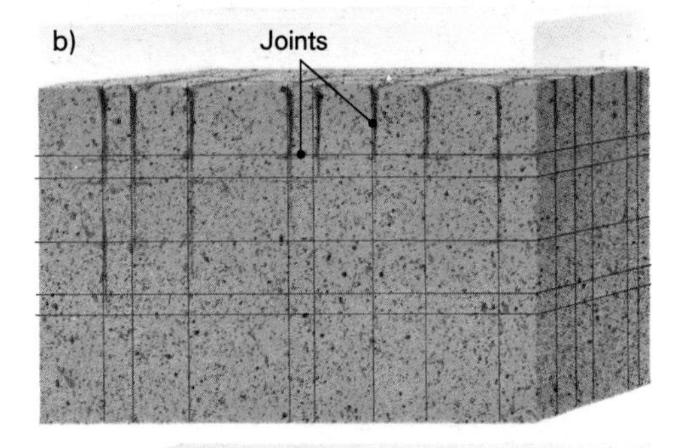

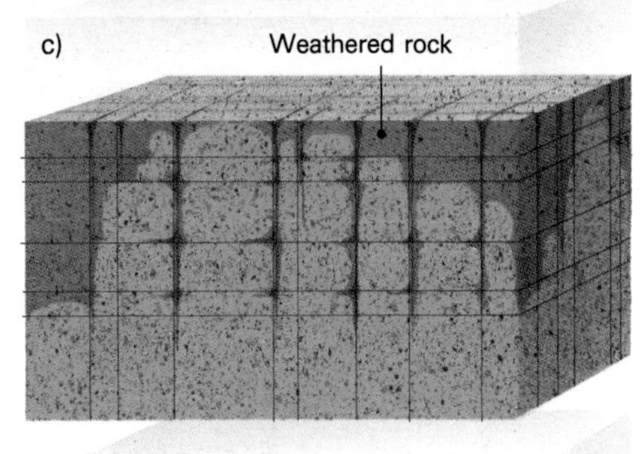

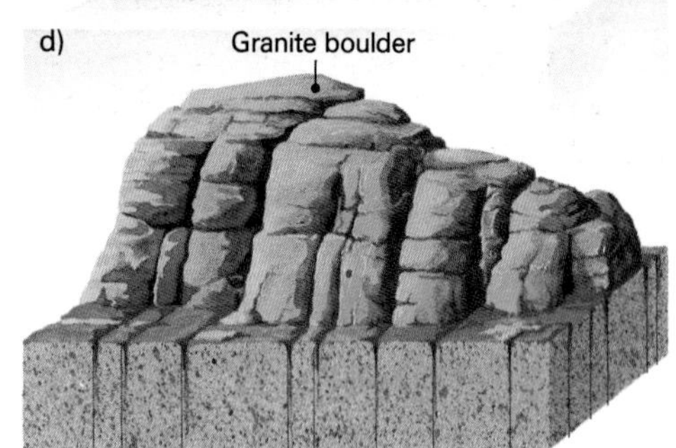

Fig. 6.2 How a tor may have been formed

1 a) Make a tracing of Fig. 6.1.
b) Label the joints and the rounded boulders.

By Professor Perry Klein
The granite is being weathered mainly by temperature changes.

It is made up of mica, felspar, and quartz (Chapter 2, p. 23). When the sun shines on the surface of the granite the crystals expand. Different minerals expand by different amounts. This happens repeatedly day after day until the crystals split apart and the rock breaks up.

Dark coloured minerals absorb more heat than pale ones which reflect heat. So pale minerals will expand less than dark ones and will split away from them.

The boulders themselves are not heated all the way through. The sun heats the surface which expands, while the inside remains cool. This leads to the outer layer splitting away from the rest of the rock. On Fig. 6.1 this is happening to the boulder on the left. This splitting away of the outer skin of the rock is called **onion weathering**.

Water seeps into the joints and gaps between the crystals. On cold nights the water may freeze. Ice takes up more space than the water from which it is made. It exerts tremendous pressure, making the joints and gaps even wider.

So the Devil's Marbles are breaking up under the effect of heat and cold. They are a fine example of **mechanical weathering**.

By Dr Crystal O'Grapher
The granite is being weathered mainly by chemical changes. It is made up of mica, felspar and quartz (Chapter 2, p. 23). The felspar is the weak link. It reacts slowly with water and carbon dioxide from the air, and becomes clay. This leads to the granite breaking up.

Impurities in the rock, like iron, are turned to rust by air and water. Rust takes up more room than the iron which increases the pressure on the crystals. Since this happens on the surface of the rock, only the outer layer splits away, exposing the next layer to the air and the weather.

This splitting away of the rock's outer skin is called **onion weathering.** On Fig. 6.1 this is happening to the boulder on the left. Its outer layers are stained rusty brown while the exposed rock beneath is white.

Salt crystals blown or washed into joints and gaps between crystals can grow and exert tremendous pressure, making the joints and gaps wider.

The Devil's Marbles are breaking up as the result of chemical action. They are a fine example of **chemical weathering.**

Fig. 6.3 Two scientific guides to the Devil's Marbles

The geologists we met in Chapter 2 (p. 32) have each produced a scientific guide to the Devil's Marbles (Fig. 6.3).

2 Hold a debate in which one side supports Professor Klein's ideas and the other side supports Dr Crystal O'Grapher.

A third scientist, Miss G. Ologist, thinks that lichen growing in the minute crevices on the surface of the rock can store water and hold it close to the surface of the rock (Chapter 5, p. 61). When parts of the lichen die, this water becomes slightly acid and can eat into the rock. She also thinks that the roots of plants growing in joints and cracks in the rock can widen them by exerting pressure on the sides.

3 a) What evidence can you find in Fig. 6.1 to support Miss G. Ologist's argument? Label this on the tracing you made for Exercise 1.
b) How convincing do you find her argument?

☆4 It is often found that different processes acting under different conditions can lead to the same result. Could all the geologists be right?

5 Rocks attacked by chemical weathering tend to be rounded, while those attacked by physical weathering tend to be more angular. Do the Devil's Marbles seem rounded or angular to you?

Weathering takes place on most rocks. For example, the carvings on Trajan's Column in Rome have been weathered through being exposed to smokey air full of sulphur dioxide from car exhausts and coal fires. Rotting animals and plants in the soil produce acids, and these and other chemicals stored in the soil can weather the rocks beneath. Indeed most of the minerals in the soil itself come from weathered rock.

Soil creep

Soil on a slope gradually moves downhill. When rain falls on the soil and sinks in, the soil expands to hold the water. The soil expands at right-angles to the slope. When the soil dries out, it contracts and shrinks. Then the surface subsides, dropping down vertically as in Fig. 6.4.

Frost can have the same effect. When water in the soil freezes, the soil is forced up at right-angles to the slope. The ice takes up more space than the water from which it is made. When the ice melts, the soil sinks down vertically in the same way as when it dries out.

This process is called **soil creep**. It is more often found in climates with frost or with frequent showers and sunny intervals. Fig. 6.5 shows some signs of soil creep, while Fig. 6.6 shows how soil scientists can measure the speed at which soil moves downhill.

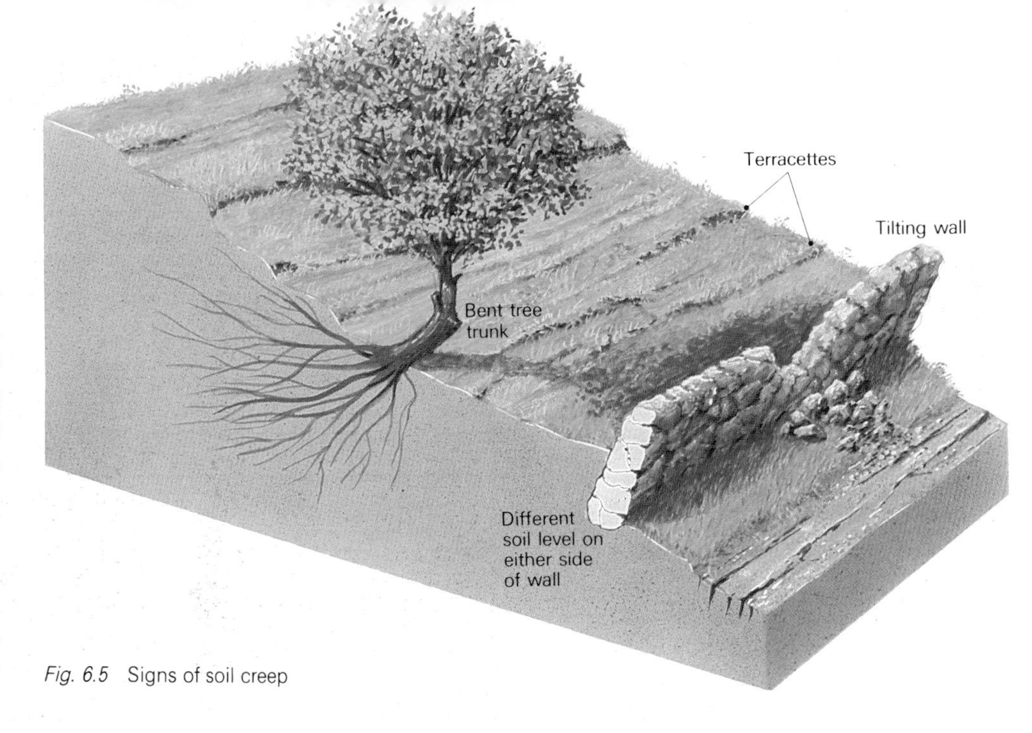

Fig. 6.5 Signs of soil creep

Fig. 6.6 Methods of measuring the speed of soil creep

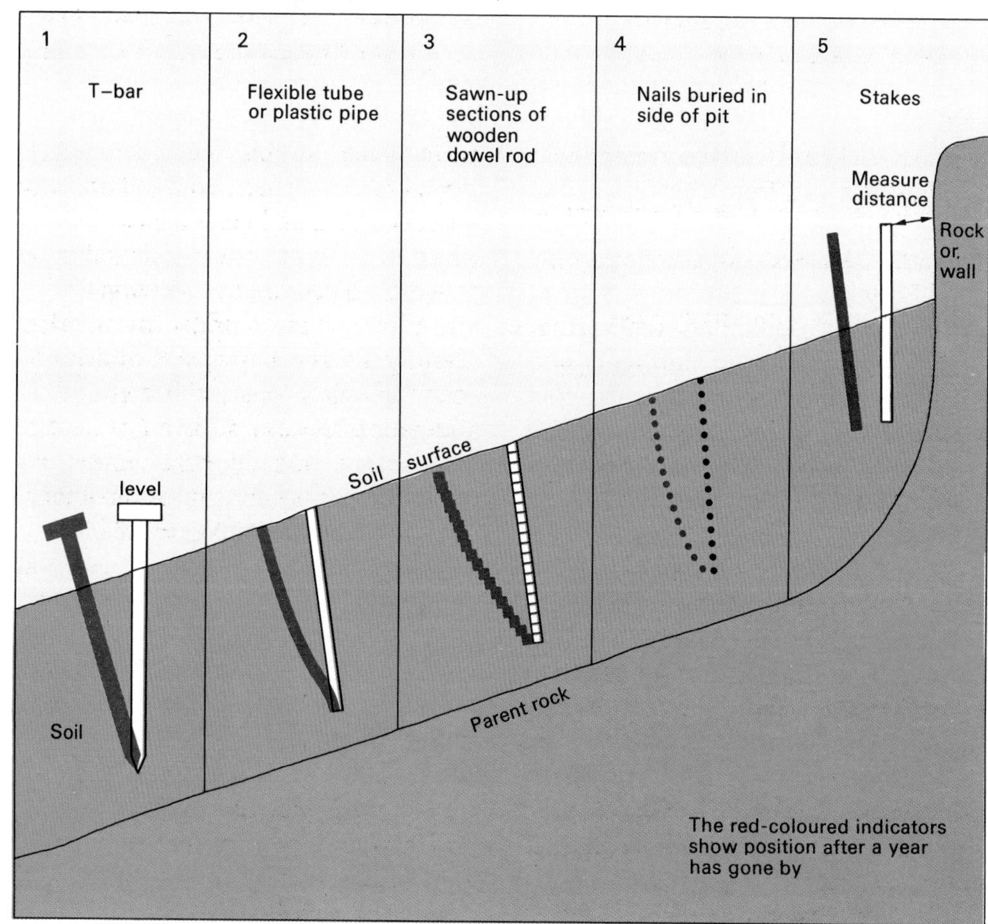

Fig. 6.4 How soil moves downhill

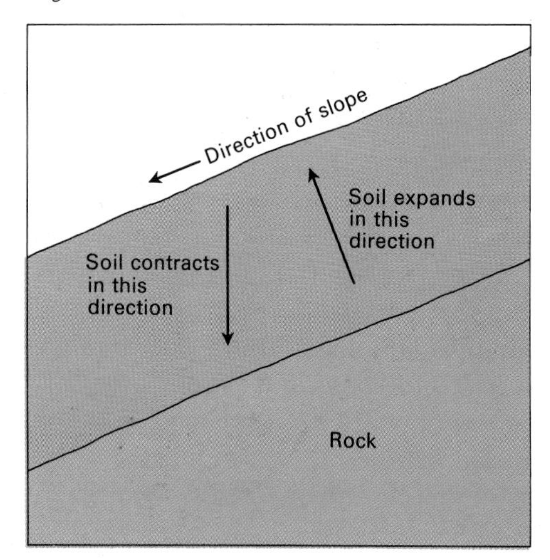

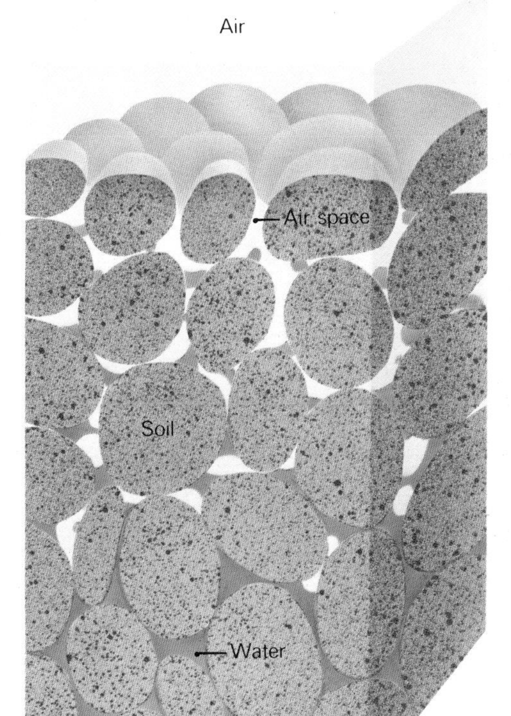

Fig. 6.7 Much enlarged section of soil

Fig. 6.8 Terracettes caused by soil creep on a steep hill slope in Derbyshire

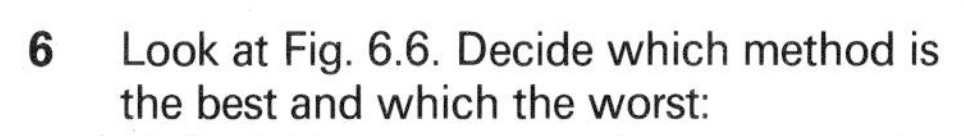

6 Look at Fig. 6.6. Decide which method is the best and which the worst:
a) for taking a number of measurements over many years.
b) for seeing how fast the soil moves at different depths.
c) for avoiding damage by cattle, vandals, etc.
d) as a compromise between the three.

7 If you have a convenient slope, you could set up an experiment based on Fig. 6.6 to be carried out over a number of years.

Fig. 6.7 shows a small section of soil with air and water trapped between the soil particles. Near the surface there is much more air than water. At the base of the soil there is no air and the spaces are entirely filled with water. In the middle layer there are small droplets of water attached to the soil particles with small pockets of air trapped between them.

Surface tension on these droplets of water helps to hold the soil together. The particles on top of the soil are only perched there and can be easily blown away. This is why dry soils suffer so much from **erosion** by the wind.

While soil creep under turf or other plants only reaches speeds of a centimetre or so a year, the removal of vegetation speeds up the movement of soil. Heavy rain on bare slopes can bounce soil particles in all directions, but they mainly move downhill.

Experiments have shown that soil can move up to 150 cm a year by being splashed by rain and the effects of this can be very serious. Rain water washing over the soil can move large amounts of soil very quickly, while water moving through the soil can carry small particles with it too. A waterlogged soil can move down a slope like porridge unless it is well bound together by roots.

8 Make some soil splash traps like those in Fig. 6.9 and carry out experiments to see how much soil is bounced out of them over a given period of time.

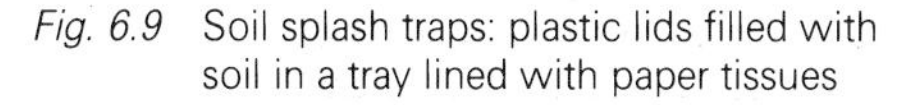

Fig. 6.9 Soil splash traps: plastic lids filled with soil in a tray lined with paper tissues

Slopes

The weight of rock on the slope in Fig. 6.10 would make the hillside collapse if it were not for the strength of the material holding it back. This may be due to the particles of weathered rock being very angular and **friction** holding the material back as the particles interlock. It may be due to the surface tension in the soil moisture making the soil cohesive.

9 a) This idea can be shown by carefully tipping into separate heaps yoghurt pots-full of dried lentils, dried peas, and dry rice, and then comparing their angles of slope.
b) Then do the same experiment when their contents have been moistened.

10 What will be the effect of the balance of strength and stress in Fig. 6.10 of:
a) water soaking into the slope increasing the weight of the upper part of the slope?
b) an earthquake?
c) building a house on top of the slope?

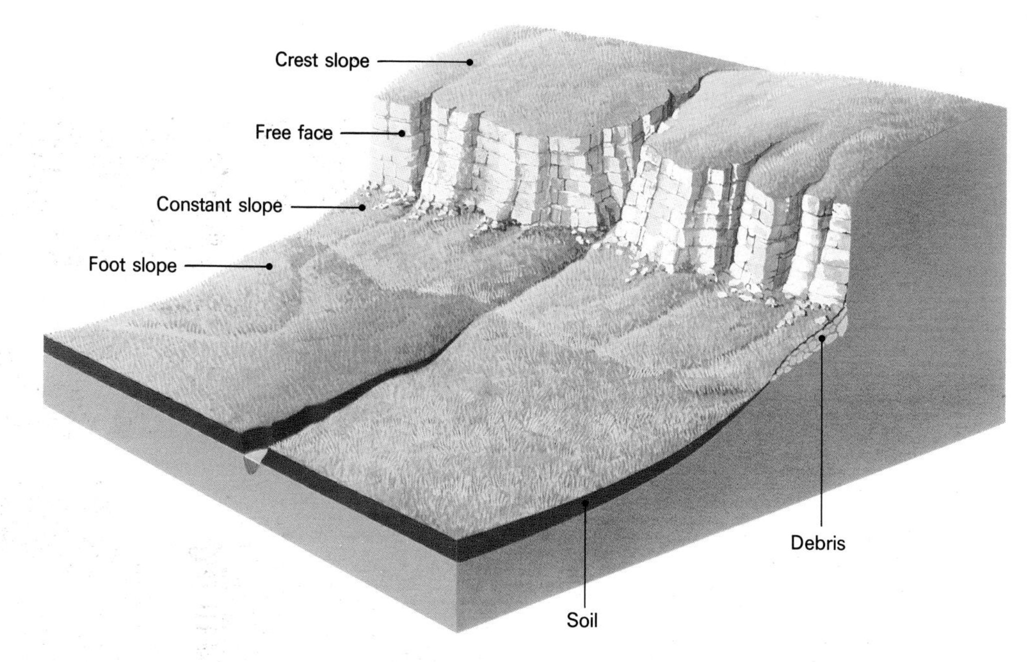

Fig. 6.11 Block diagram showing the different parts of a hillside

11 Read the extract from the *Daily Mirror* of Monday, 6 March 1978 (Fig. 6.12). Which of the causes mentioned in Exercise 10 do you think could have been the reason for the landslip?

Fig. 6.11 shows a block diagram of a hillside.
i) At the top or crest of the hill there is the gently curving **crest slope**.
ii) Below this there may be a steep slope or even a cliff of solid rock—the **free face**.
iii) Below this is a section of straight slope called the **constant slope**, which is made up of material which has been washed or has fallen from the free face above.
iv) Below the constant slope is the **foot slope**, a gently inclined section covered in soil.

12 a) Look at any nearby slopes and hillsides and at any photographs you can find. Figs. 2.4, 2.8, 2.20, and 2.26 will be useful here.
b) Sketch some of these and see if you can mark on the various slopes of a hillside shown in Fig. 6.11.
c) Notice that seeing the true shape of the slope often depends on the angle from which you are looking.

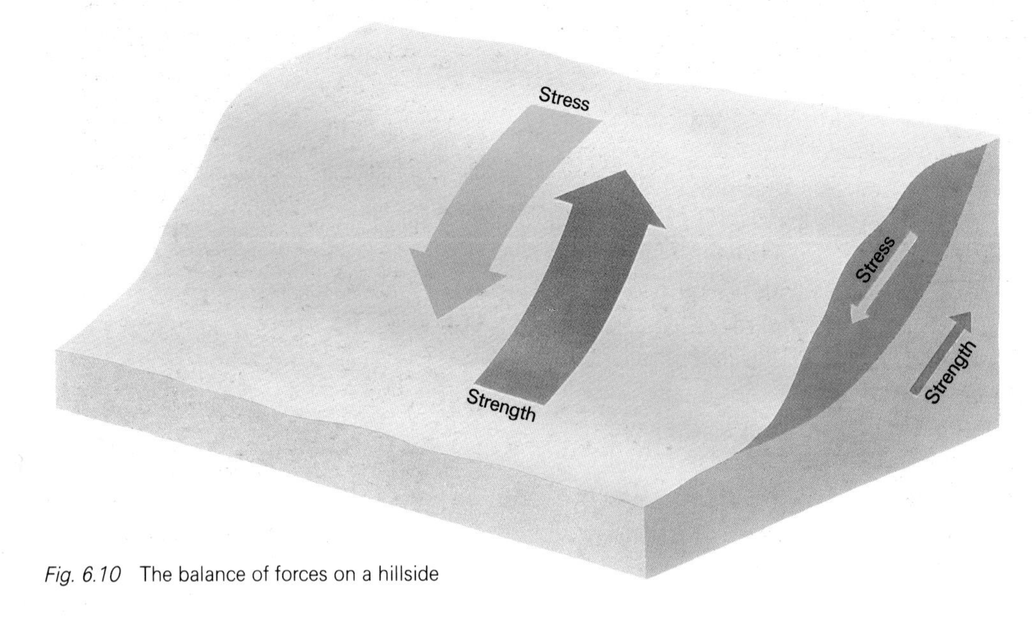

Fig. 6.10 The balance of forces on a hillside

Fig. 6.12 (right) A very sad newspaper story

CRUMBLE COTTAGE

Landslide wrecks cliff homes

By Alan Gordon

The Watsons are moving house ... because their house is on the move.

The front porch isn't where it used to be. The floors tilt crazily—and so does the only road out.

They all collapsed under the onslaught of a massive landslide which hit the southern tip of the Isle of Wight at the weekend.

And suddenly the Watsons' home, Cliff Cottage, was reduced to Crumble Cottage.

Move

Yesterday, all they could do was pick up the pieces ... and move on.

Sisters Louise Watson, thirteen, and Emma, ten, helped to carry out the furniture for their father Rex, a garage owner. Louise said: 'We are so sad.'

There were other refugees, too, from the big crack-up at the remote beauty spot of Blackgang.

A dozen people lived in the six houses affected. Radio dealer Richard Young put his losses at £50 000. A holiday cottage which he lets was flattened and his home badly damaged.

He said last night: 'It was amazing to see the cottage go down. Now it looks like some giant bonfire waiting to be lit.' Mr Young will have to bear the damage himself. Insurance companies refused to carry the risk—because the area is subject to subsidence.

Local people said that masses of blue underclay soaked by heavy rains cause the erosions.

More havoc could be on the way. Police and coastguards say that another 25 acres are slowly on the move.

End of the road for the house that slipped away

Picture **Alisdair Macdonald**

Erosion by the sea

Fig. 6.13 shows the cliffs and houses collapsing at Pakefield, south of Lowestoft. These cliffs are made of soft, crumbling clay and offer very little resistance to the sea. Yet far more solid rock cliffs give way to the sea's attack. How does sea water erode hard rock cliffs?

Firstly, the salt sea spray in the air helps to weather the rocks. Salt crystals grow and expand in cracks and crannies to cause the rock to crumble. Secondly, the waves trap air in between themselves and the cliff face, which exerts huge pressures on cracks in the cliff. Thirdly, the sea scours against the foot of the cliff, often armed with pebbles and rocks, causing **abrasion** of the cliff. A **wave-cut notch** may appear which undermines the cliff and causes it to fall.

Usually there is a kind of balance kept in the way the sea erodes. If the sea is successful in eroding a cliff, rocks and boulders fall to the foot of the cliff forming a beach. This acts as a barrier protecting the cliff from further erosion, though it is still exposed to weathering by the salty spray.

Fig. 6.13 Buildings damaged by erosion in 1906

Fig. 6.14 Handfast Point, Dorset: a chalk headland eroded by the sea

If a cliff retreats rapidly, a **wave-cut bench** will be left off shore. This keeps the water shallow and robs the waves of much of their strength.

Inlets cut by the sea along lines of weakness in the rock will often be protected from wave erosion by the headlands on either side. In an inlet a beach may form, protecting the shore still further from attacks by the sea. The headlands will receive the full force of the waves.

13 a) Make a copy of the flow diagram in Fig. 6.17.
b) Which of the pictures shows what is happening in box A, box B, box C?

14 a) Make a plasticine model at one end of a seed tray of the headlands shown in Fig. 6.15. Use more plasticine to fill any drainage holes in the tray.
b) Half fill the seed tray with water to act as the sea.
c) Use a ruler parallel with the end of the tray to make little waves.
d) Draw a sketch plan of the tray and show the path taken by the waves approaching the shoreline.
e) Repeat the exercise with the ruler at different angles to the coastline in order to bring waves in from different directions.
f) You can use this method to make models of other stretches of coastline. If you put a half-inch layer of sand in the bottom of the tray, you can see what happens to material washed from the cliff.

Fig. 6.15 A seed tray wave tank

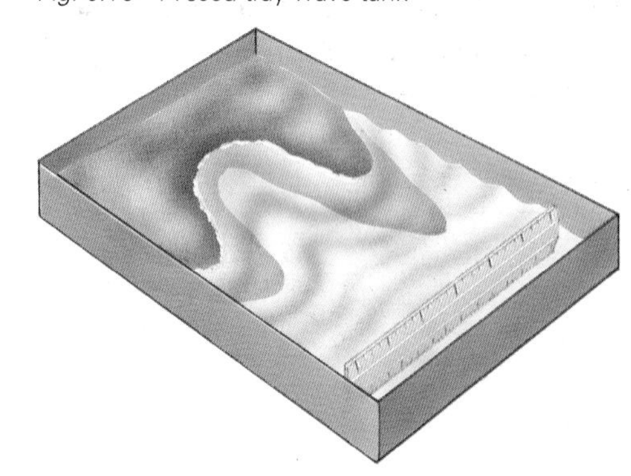

When a wave approaches a headland, it will often change its course until the wave line runs parallel with the coastline. As it meets the shallow sea floor near the coast, the wave will swing round until it faces the shore head on. This bending round of waves to meet the coast is called **wave refraction**.

As a result the sides of headlands are prone to attack from waves. Eventually **caves** may be cut in both sides of a cliffed headland to form an **arch**. In time this arch will collapse and leave an isolated **stack** standing in the sea (Fig. 6.14).

15 Make a sketch of Fig. 6.14. Label the following features: stack, headland, arch. Find out the name of the stacks.

Not all coastal cliffs are under attack by waves. Hengistbury Head in Dorset (Fig. 6.16) looks at first sight as though the sea is successfully eroding it away. Yet a closer look shows that the foot of the cliff is well above the level of high tide and in places grass is growing there.

Instead the cliffs have been attacked in three other ways. Firstly, there has been quarrying, yet another example of people eroding the land (Chapter 2, p. 28). Secondly, millions of tourists have trampled the grass at the top of the cliffs to produce bare soil which is very liable to erosion. Thirdly, springs of water in the cliff face wash material away. At times they cause the cliff to collapse.

16 On a tracing of Fig. 6.16 draw in where the quarrying, the trampling, and the natural springs have eroded the cliff.

Fig. 6.16 Three types of erosion at Hengistbury Head

Fig. 6.17 Erosion by the sea: flow diagram for Exercise 13

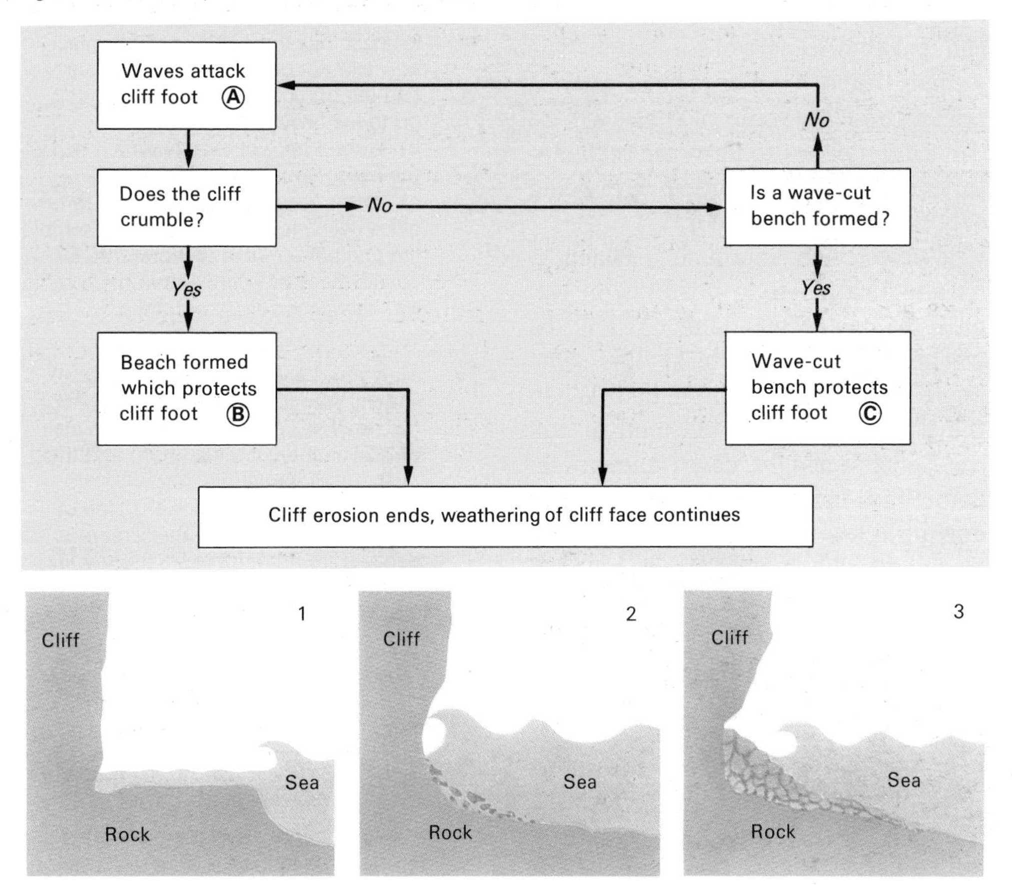

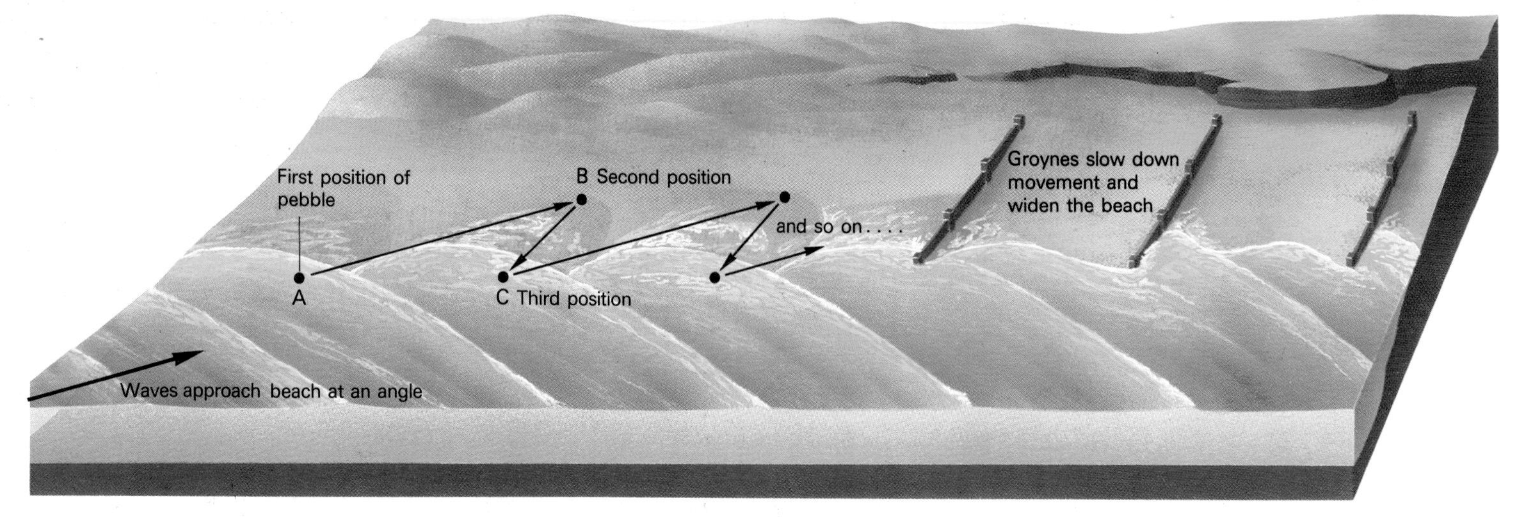

Fig. 6.18 How longshore drift can move pebbles along a beach

Transport by the sea

Once a rock has fallen to the foot of a cliff it will form part of the beach. The same is true of sand or pebbles washed out of the cliff. Some of these will be carried out to sea to form the sedimentary rocks of the future by being deposited on the sea floor.

Waves often approach a beach at an angle rather than head on. In Fig. 6.18 a pebble at point A will be washed up the beach at the same angle as the wave approached. When the wave retreats, it will tend to pull the pebble straight down the beach. Sometimes the pebble will roll straight down the beach under its own weight.

Eventually it will reach point C ready for the next wave to send it up the beach again. This movement of pebbles or sand along a beach is called **longshore drift.**

17 Use the seed tray wave tank from Exercise 14 to make a model of longshore drift at work. This time use sand or beads to make a straight beach at an angle to the side of the tray.

18 Try experiments with coloured pebbles to see if longshore drift is taking place.
a) Paint the pebbles a bright colour and put numbers on them before you go to the beach.
b) Make a note of exactly which part of the beach you leave each pebble.
c) This is fun to do, but it needs to take place over a long time if it is to resemble the true pattern of longshore drift. Of course many of your painted pebbles will be washed away or buried.

19 a) If your beach has groynes or breakwaters, you can compare the height of the beach on each side of the groynes to give an idea of the amount and the direction of longshore drift.
b) Is the beach higher on the side of the groyne from which the sand or pebbles are drifting? The pattern will often change from day to day, particularly in stormy weather.

Longshore drift can cause many problems. Entire holiday beaches can be washed away, often in a single year. Some resorts bring sand in lorry loads to 'feed' the beach. Palm Beach, Florida, in the USA has a system of offshore dredgers which pump the sand back to the beach along pipelines.

Beaches protecting the foot of cliffs are often themselves protected by groynes and sea defences. Harbours can be silted up by sand deposited by the sea.

The spit game

Fig. 6.19 shows an area of land with a shingle beach on the south-west side. Many of the flint pebbles on this beach have been eroded from a cliff to the west of the beach. They are being carried south-eastwards by longshore drift. The object of the game is to build a line of shingle which looks like Hurst Castle spit, the sandy area on Fig. 6.19.

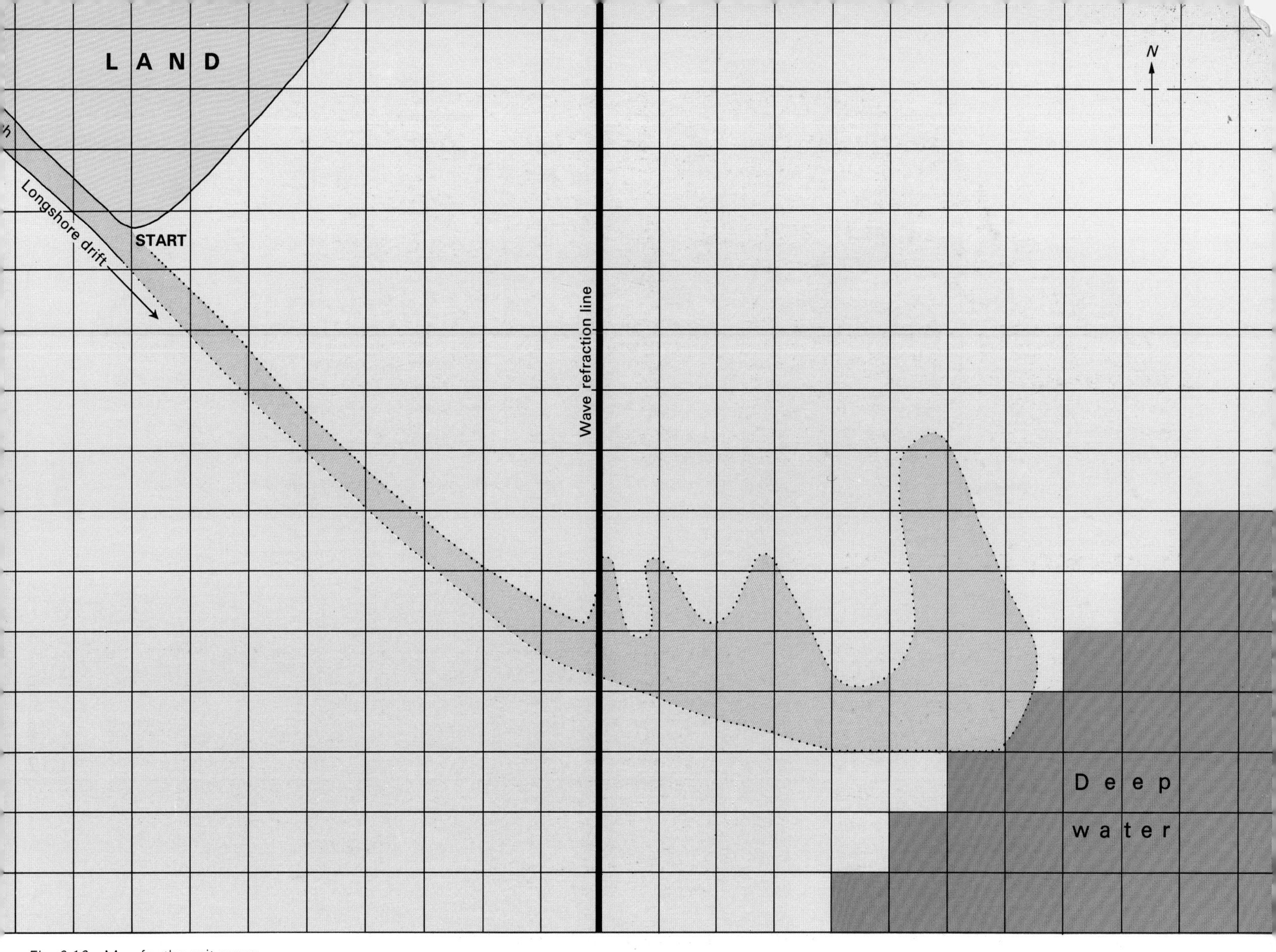

Fig. 6.19 Map for the spit game

1 Lay a sheet of tracing-paper over Fig. 6.19 and mark in the corners of the map and the outline of the land.

2 You need a pack of 50 cards (Fig. 6.20) as follows: 20 west wind cards, 8 south-west wind, 6 south-west gale, 10 north-east wind, and 6 south wind. From all other wind directions the area of sea is protected by the land on your map and by the Isle of Wight to the south-east.

3 Shade in the start square. Then work through the pack, which should be well shuffled, following the instructions on each card. Shade the squares in pencil so that you can rub out easily when necessary.

4 The south-eastern corner of the map has deep water so that you will need three west wind cards in a row to extend the beach into any deep water squares. The other cards count as usual.

5 The effect of wave refraction is shown by the vertical line. East of this line, wave refraction makes the spit veer round to the east. So west winds build the beach only in an easterly direction.

6 At the end of the game compare your results with the actual area of Hurst Castle spit. To work out your score give yourself a penalty point for each square you have shaded which does not include any part of the spit. Also give yourself a penalty point for each square within the actual area of the spit that you have not shaded. The winner will have the lowest score.

Fig. 6.20 Wind cards for the spit game

Fig. 6.21 Hurst Castle spit from the south-east at high tide

20 After you have played the spit game:
a) Draw a circle in the middle of a page and label it Hurst Castle spit.
b) Draw four arrows approaching this circle to represent winds from the west, the south-west, the south, and the north-east.
c) Use the wind cards in Fig. 6.20 to help you label each of the arrows to sum up the effect of winds from these directions.

West of Hurst Castle spit groynes have been built to stop the shingle being moved away. This builds up the beach and protects the cliffs from erosion. But it also means that the spit could disappear. The spit protects marsh and farmland which has been built up in its shelter to the north. If the spit disappears, then erosion could set in, removing large areas of land to the north and north-east.

21 Do you think that the local council would do best to:
a) remove the groynes and allow the eroding cliffs to provide more shingle?
b) tip hundreds of lorryloads of shingle to feed the beach just west of Hurst Castle spit (start square on Fig. 6.19)?
c) leave things as they are and allow the spit to disappear?

22 Find out all you can about spits on the coastline of Britain. Find Orford Ness, Spurn Head and other spits in your atlas and see if you can work out how they might have formed.

River erosion

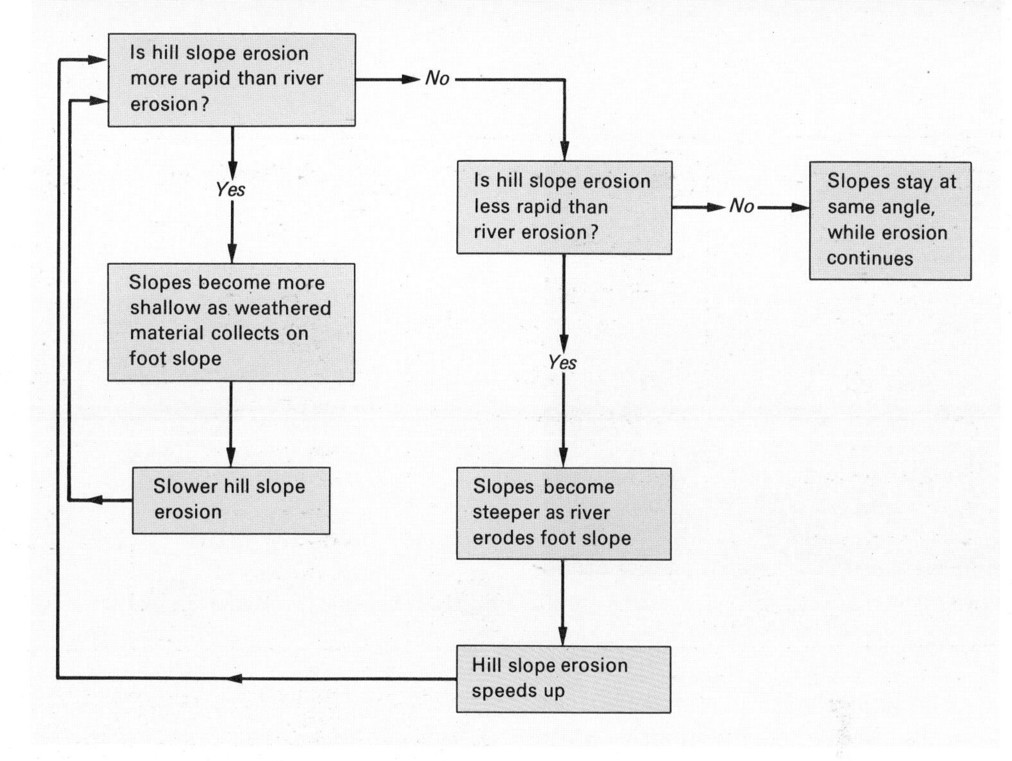

Fig. 6.22 Flow diagram to show the connection between hill slope and river erosion

Many people in Britain live a long way from the sea so it is often more convenient to study erosion in a river or stream. A river erodes the rocks exposed in its channel in four main ways.

i) Rocks which are broken or jointed are sometimes lifted and carried away by the sheer force of the moving water; this is called **hydraulic action**.

ii) The river water may also attack the bed chemically by **corrosion**.

iii) The most effective form of erosion seems to be where the river uses its load to batter its bed; this can be called **abrasion** or sometimes 'corrasion'.

iv) This battering has the effect of wearing down the load carried by the river. This is called **attrition**.

It is very difficult to see these processes at work because erosion is most effective when rivers are in flood.

In a section of countryside drained by a river, weathered rock in the form of boulders, pebbles, sand or silt is moved from the slopes down to the footslope. The material gathers near the edge of the river ready to be removed by the river itself. The river's main job (Chapter 4, p. 52) is to transport this material further down its course and eventually out to sea.

If the river erodes its bed more rapidly than the rate of erosion on the valley sides, then the hill slopes become much steeper (Fig. 6.22). This greater steepness of the valley sides increases their rate of erosion until both the hillsides and the river bed are being eroded at the same rate. Then, the angle of the slopes remains constant while erosion continues. This idea is shown in **cross-section** in Fig. 6.23.

Fig. 6.22 also shows that if the valley sides are being eroded more rapidly than the river bed, the slopes will become shallower. This has the effect of slowing down the rate of hillside erosion until both the river bed and the hillsides are in balance, eroding at the same rate. Once again the valley sides will then stay at this constant angle even though erosion continues.

23 Using the text and Fig. 6.22 draw a cross-section similar to Fig. 6.23. This time show what happens if the river erodes its bed less rapidly than the erosion on the valley sides.

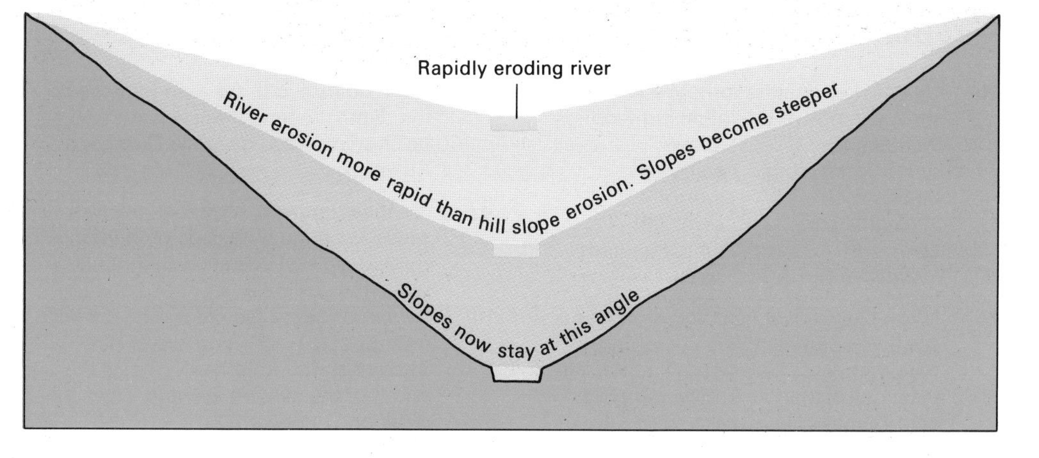

Fig. 6.23 Three cross-sections of the same valley showing the effect of the river bed eroding at a faster rate than the valley sides

Fig. 6.24 A small delta forming at the south end of Lake Derwent, Cumbria

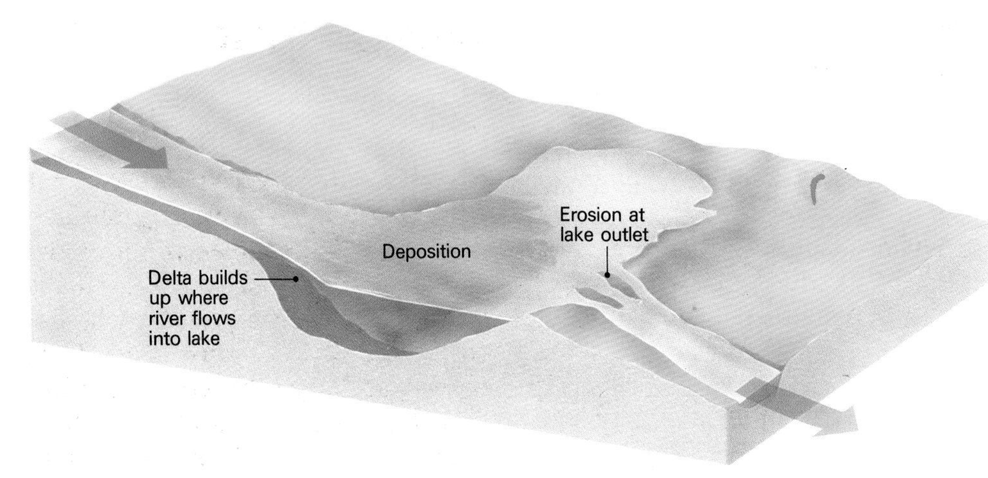

Fig. 6.25 Stages in the removal of a lake from a river's course

A **long profile** is a section of a river from its source to its mouth. Most rivers flow into the sea. They cannot cut down any lower than sea-level otherwise they would have to flow uphill to the sea. So for most rivers sea-level is **base-level**. On a smaller scale a lake can act as base-level for the streams that flow into it.

Compared to the millions of years of geological time shown in Fig. 1.24, p. 18, lakes do not last for very long. A lake traps silt and sediment brought down as a river's load. Slowly these silts will build up as a delta which grows out from where the stream enters the lake (Fig. 6.24).

Fig. 6.25 shows that where a lake drains downstream into a river, erosion will take place until the whole lake is drained and thus disappears. The river then flows between terraces of old lake beds.

24 Fig. 6.26 shows the stages in the removal of a waterfall by river erosion. Explain in your own words what happens.

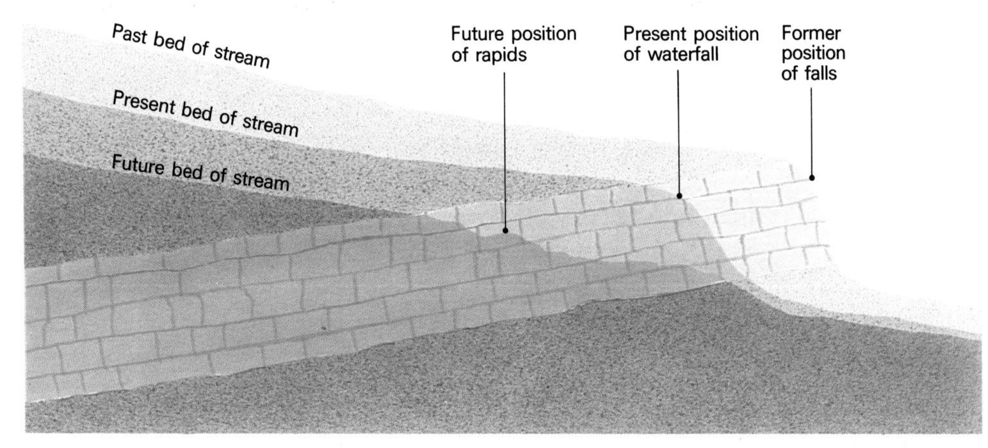

Fig. 6.26 Stages in the removal of a waterfall by river erosion

Summary exercises

25 Use the last chapter to help you explain the meanings of the following words:

weathering	joints
onion weathering	soil creep
mechanical weathering	erosion
	abrasion
chemical weathering	longshore drift
	spit

26 If you have a river near you, visit it and see if you can identify some of the following features: waterfall, rapid, lake, meander, estuary, tributary, gauging post (it looks like a giant-sized ruler stuck on its end in the river to measure the depth). Some of these features may be shown on a large-scale Ordnance Survey map.

27 Meandering rivers have their deepest sections on the outer side of the bend; these deep sections are called pools, while the shallower sections between the bends are called riffles. The distance from pool to pool along a river has often been found to be five times the width of the river on average. Does this work on your river?

7 Snow and ice

Erosion by ice

No one is sure why ice ages occur, but lots of reasons have been suggested.

It may be that great storms on the sun's surface called sunspots reduce the amount of heat given out by the sun. Huge clouds of dust or gas from outer space may drift between the sun and the earth, reducing the heat reaching the earth's surface. Great clouds of volcanic dust or ash may also reduce the heat reaching the earth's surface.

The earth may wobble on its axis, stopping the northern or southern half of the earth from pointing quite so much towards the sun in July or January. This could result in colder summers than usual and could be enough to start another ice age.

Nor are we certain how long it would take for another ice age to start. Some people think that it would take 15–30 thousand years for the ice sheets to spread gradually out from the poles. Others think that one very cold summer would be enough to start a vicious circle of cooling which could lead to an ice age.

Fig. 7.1 Snow in the shaded hollow of Ben Nevis (the mountain on the right) tends to last throughout the year

1 a) Draw up a list of the different possible causes of an ice age. Then add two columns, one for slow, and the other for rapid, starts to an ice age. Add ticks for each possible cause depending on whether you think it is most likely to be rapid or slow.
b) Write a sentence or two below your table to sum up your conclusions.

In any north-facing hollow snow will tend to linger on into the summer. It was only in 1976 that the snow melted completely on Ben Nevis in the Grampian Highlands in Scotland. Gradually, year by year, the snow piles up. The air is squeezed out from the lower layers by the weight of the snow above. The bottom of the snow then turns to ice.

If this process continues, a great mass of ice will begin to grow in the most sheltered parts of the hollow, close to the steepest part of the north-facing slope. As the weight of this great mass of ice pushes down at the back, the ice in front gradually slides out of the hollow.

As the ice moves, a number of things happen. Have you ever put your hand into the ice-box of the refrigerator to find that it has stuck to the ice-tray? The ice at the back of the hollow sticks to the ground it rests on and **plucks** away the rock, soil, and stones. These are ground down against the floor of the hollow and used as tools to **scour** the hollow deeper. As the ice flows out of the hollow, it **smooths** the **lip** of the hollow, using the stones it carries as its load like sandpaper (Fig. 7.2).

Fig. 7.3 shows the deep armchair-shaped hollow that is left when the ice has melted. It is called a **corrie** and the type of glacier that formed it is therefore called a **corrie glacier**. It often fills the valley below the corrie with a great thickness of ice. The weight of this, and the rocks and stones it carries, will deepen the valley slowly until it becomes a deep glacial trough.

In many river valleys the river flows between spurs of land as shown in Fig. 7.4. When such a valley is filled with a glacier, the ice tends to straighten the valley leaving **truncated spurs**. The floors of any side valleys are then well above the level of the main **glacial trough** (Fig. 7.5).

Fig. 7.6 shows a mountain area in Britain as it might have been before the Ice Age. Rivers flowed out in all directions from the central ridge. When the Ice Age arrived, corrie glaciers formed in the hollows and round the springs at the heads of the rivers. Ice flowed as glaciers down the main valleys, deepening them into glacial troughs. Where two glaciers flowed together, the valley downstream was gouged out deeper and wider.

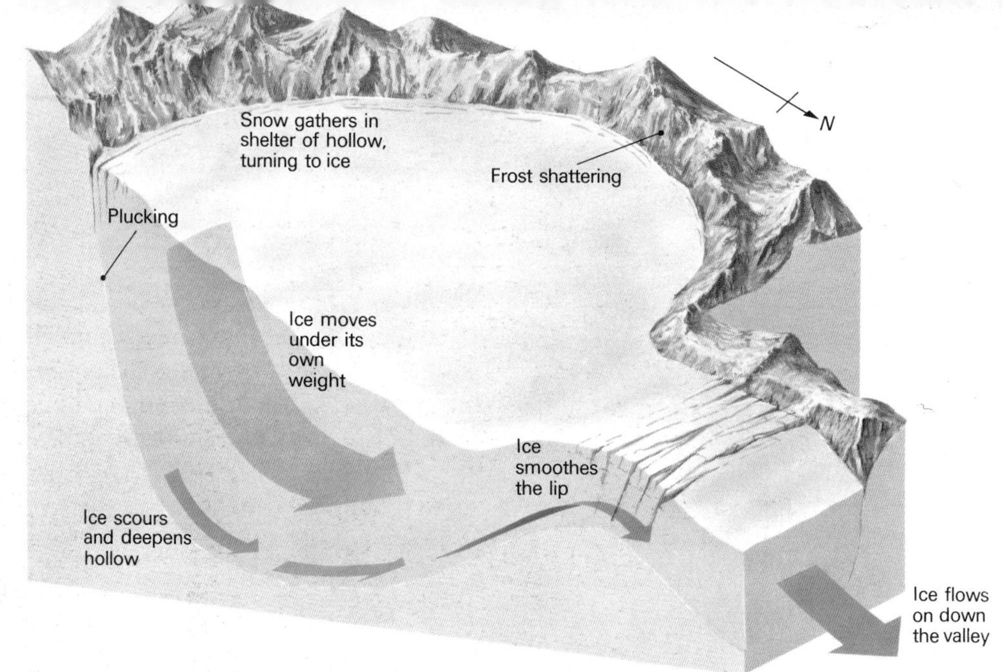

Fig. 7.2 A section through a corrie glacier

Fig. 7.3 A huge corrie with a lake below the summit of Helvellyn in Cumbria

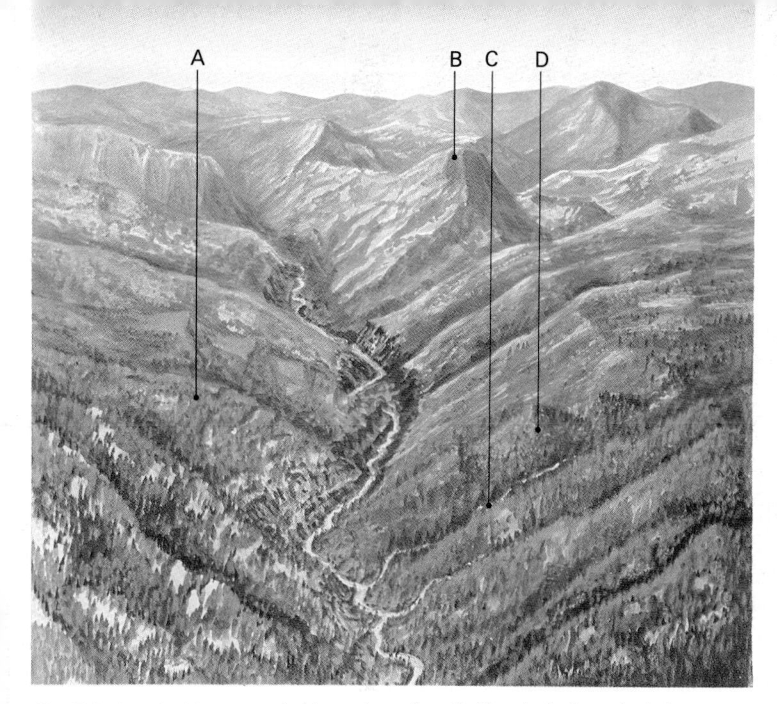

Fig. 7.4 Interlocking spurs in Yosemite valley, California, before glaciation

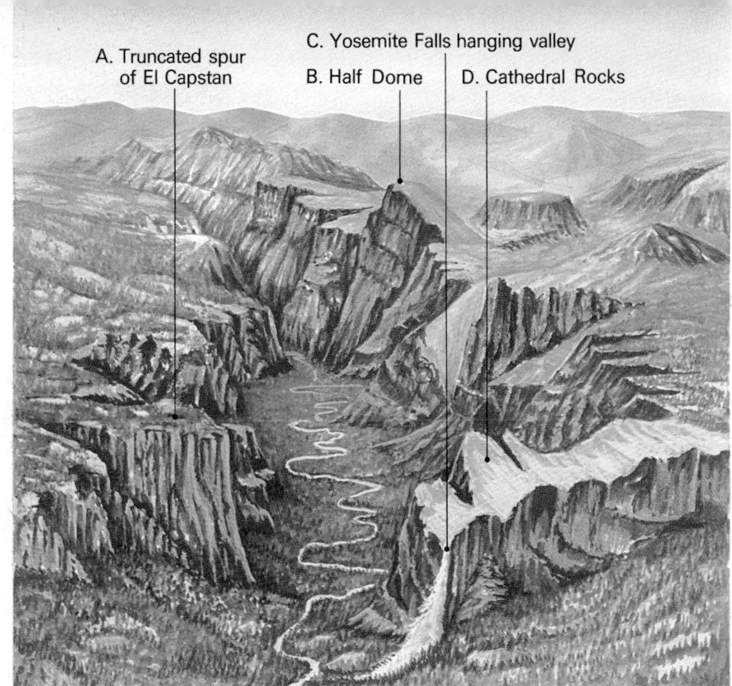

Fig. 7.5 Yosemite valley after glaciation: a glacial trough with truncated spurs

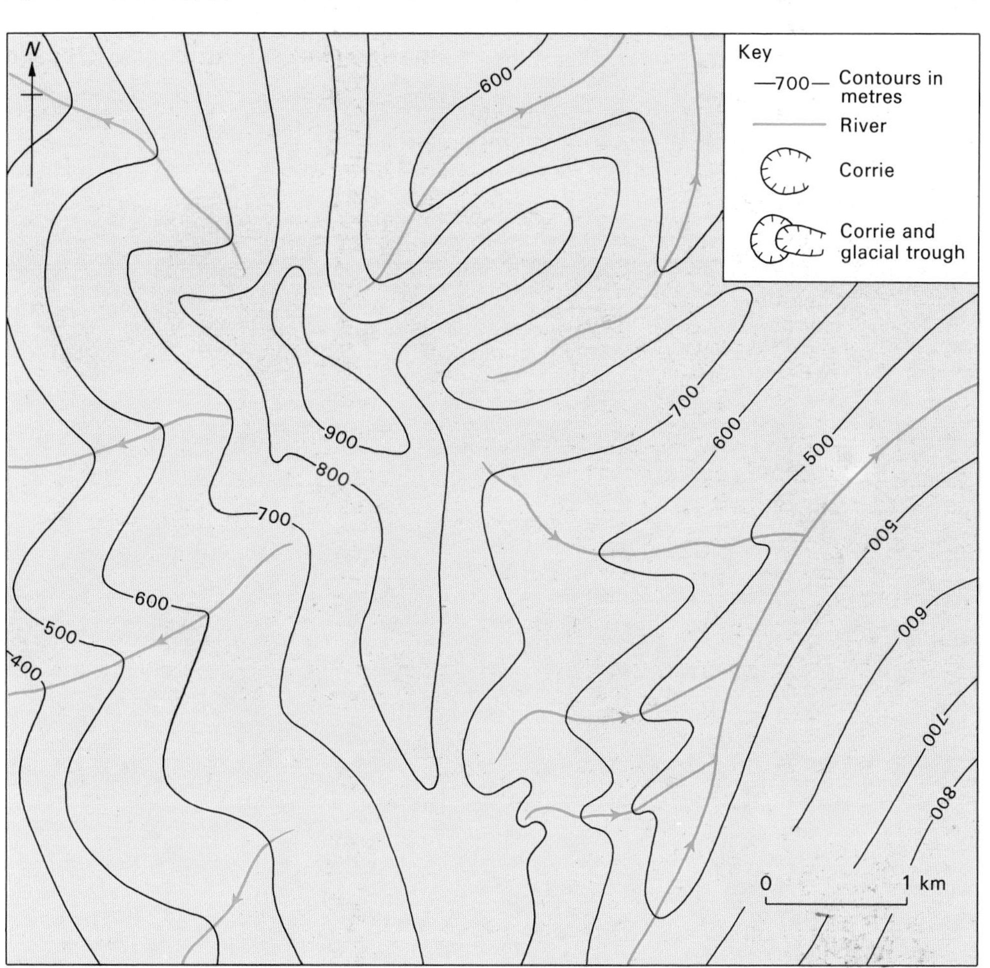

Fig. 7.6 A highland area of Britain

2 On a copy of Fig. 7.6 shade very lightly the areas between the contours using the following key:

over 900 m	grey
800–900 m	purple
700–800 m	brown
600–700 m	yellow
less than 600 m	green

3 a) On tracing paper laid over your copy of Fig. 7 6 plot corries at the head of the valleys, using the symbol shown in the key.
b) Add glacial troughs to your overlay, using the symbol shown in the key. The width of the valleys should be about half the diameter of the corries at their head.
c) Compare your map with Sheet 90 of the Ordnance Survey 1:50 000 or the Tourist Edition map showing Helvellyn. Make a note of the similarities. Try to explain any differences.

What the ice leaves behind

The rocks and soil picked up by a glacier form its load or **moraine**. Material that drops onto the glacier from the valley sides is called **lateral moraine**, while the material deposited at the end or **snout** of the glacier is called an **end moraine**.

When the ice has melted, these features can sometimes still be seen. They may dam a valley to form a lake. If a highland area has been glaciated, it often has many lakes. The Lake District of Cumbria is a very useful illustration of this (Fig. 7.8).

Fig. 7.7 End moraine of a glacier viewed across a frozen lake in Switzerland

4 The labels in Fig. 7.9 have been mixed up. Make a copy of the drawings with their correct labels.

Ice sheets cover very large areas and they often scour away all the soil from the places they cover. The ice

Fig. 7.8 Wastwater: a lake in a deep valley left by a glacier

Fig. 7.9 Four types of lake for Exercise 4

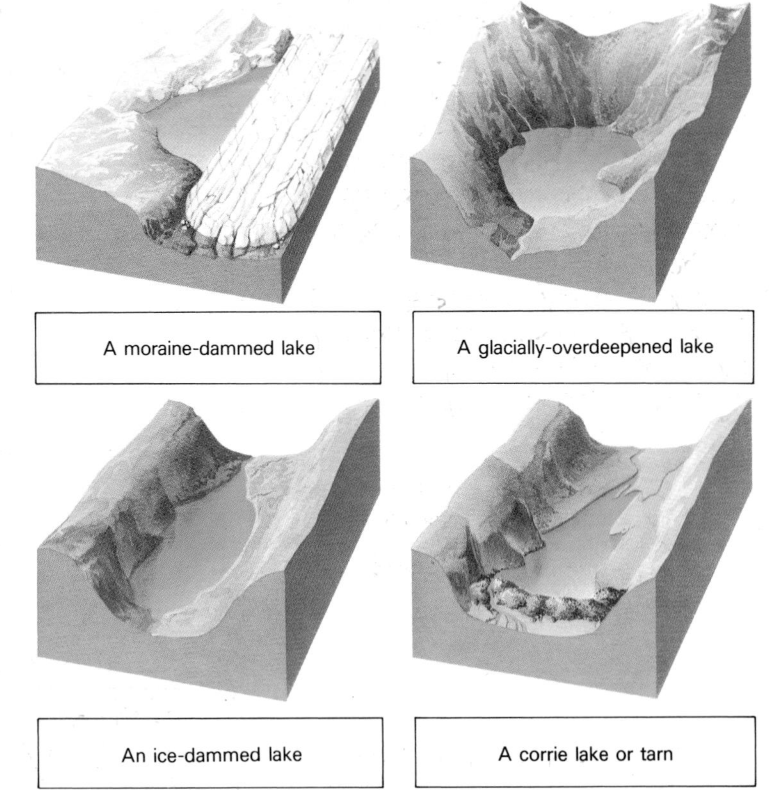

moves slowly out from the centre of the icefield. It can fill in valleys and make the drainage pattern of an area very complicated.

Towards the edge of the ice sheet much of the material scoured up from elsewhere is smeared over the rocks to form a thick layer of **till**. This takes the form of clay with large boulders carried by the ice from long distances away. It is called **boulder clay** in the English Midlands and East Anglia.

Occasionally a mass of ice can be buried in the boulder clay. When this melts, it forms a rounded depression called a 'kettle hole'. This will fill with water to form a round pond. Not all round ponds in clay are kettle holes. Many may have been dug out by man as clay or **marl pits**.

The boulders or rocks which have been moved huge distances by the ice are often totally different from the rock on which they now rest. Such rocks are called **glacial erratics** (Fig. 7.10). Many are found in the Midlands and Yorkshire; the smaller ones were often used in the days of horse transport as mounting blocks to help people get on to their horses.

These erratic rocks are useful because they help to show the direction of flow of the ice. The pattern of ice flow in Britain is shown in Fig. 7.11.

5 a) Use an atlas to help you to make a list of the towns along the southern limit reached by the ice in Fig. 7.11.
b) Was your home area glaciated?

Fig. 7.10 A glacial erratic at Norber Rocks near Austwick, North Yorkshire

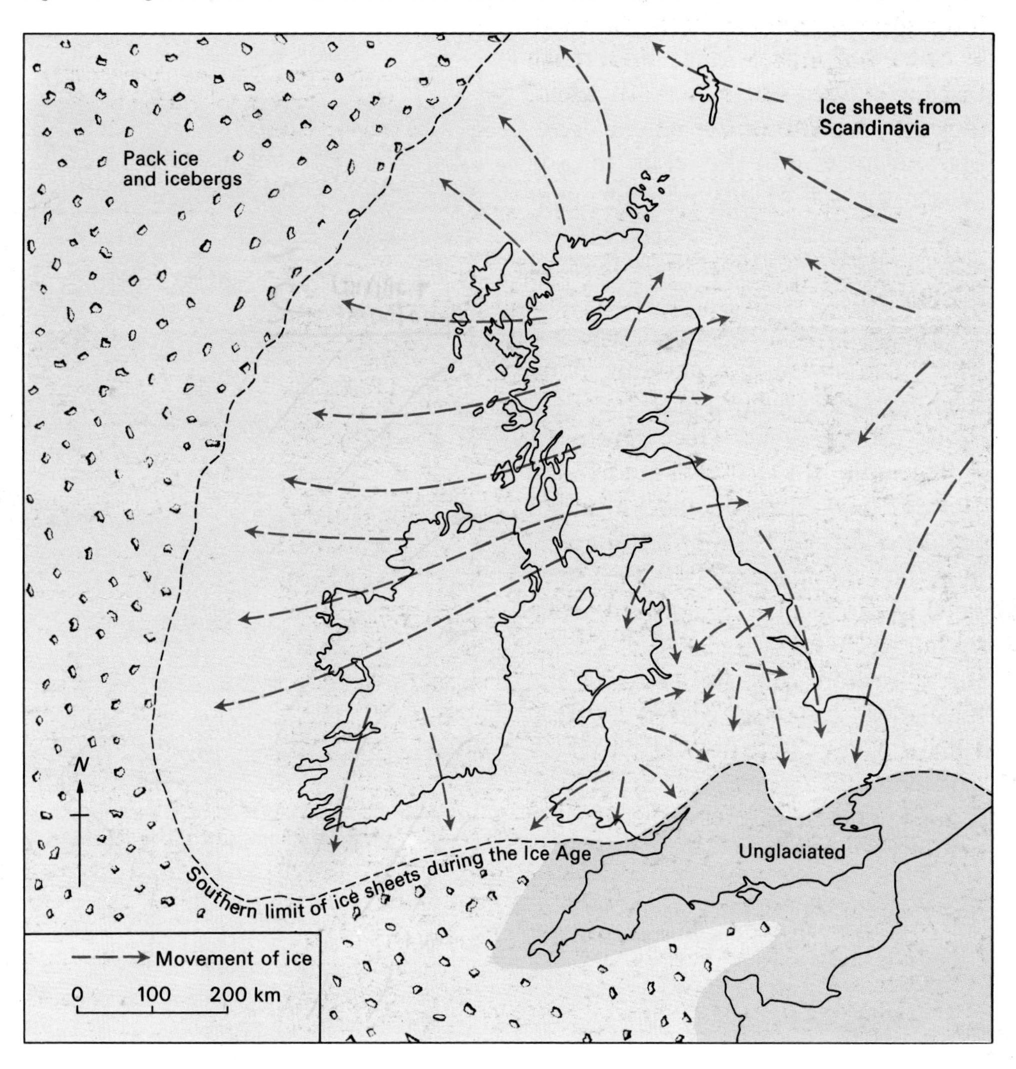

Fig. 7.11 Pattern of ice flow in Britain during the Ice Age

Beyond the ice sheet

The areas close to the edges of the ice were still very cold, and suffered from what are called **periglacial** conditions. The soil and the water it contained were permanently frozen. The ice in the soil occupied a greater space than the water from which it was formed. It often heaved up the surface of the soil, just as frost will sometimes do to concrete, making gates and doors difficult to open.

Frost often shattered any boulders or stones on the surface and today these shattered fragments lie where they fell. Sometimes the regular freezing and melting of the ice in the soil caused the soil to heave up by night and to sink again by day. The sides of gravel pits often show the contorted layers which result from this (Fig. 7.13). Loose stones on the surface were often sorted into patterns as on the Stiperstones ridge (Fig. 7.12).

On steep slopes this movement would make the soil move downhill. Many of the U-shaped valleys found in highland areas may owe their shape to material slumping down the steep sides of the glacial trough and filling up the valley floor.

Meltwater from ice caps on the hills flowed down and carved out steep valleys. Many of the dry valleys which exist on chalk or limestone areas were formed by such meltwater flowing over frozen soil.

6 All areas of Britain show periglacial features. Try to find out what periglacial features exist in your home area.

Fig. 7.12 Geography students studying the patterned ground on the Stiperstones in Shropshire

Fig. 7.13 Layers of gravel on Hengistbury Head contorted by frost heave during the Ice Age

Summary exercises

7 Use the last chapter to help you to explain the meanings of the following words:

- ice age
- corrie
- truncated spur
- glacial trough
- lateral moraine
- end moraine
- corrie lake
- kettle hole
- glacial erratic
- till
- periglacial features

8 a) Under periglacial conditions, in a valley running from east to west, which side of the valley is more likely to have ice melting during the day and freezing again during the night?
b) Which side of the valley will become steeper as a result of material slumping down the slopes?

8
Wind

Land and sea breezes

Why does wind blow? Where does it come from and where does it go to?

In order to answer these questions it is best to start with a gentle breeze. There are three important points to remember:

i) Land heats up and cools down more quickly than water.
ii) Warm air rises and cool air sinks.
iii) Air is heated or cooled by the land or sea below it.

Fig. 8.1 shows a beach. In the morning the land heats up more rapidly than the sea and warms the air above it. The warmed air begins to rise which lowers the pressure above the land. So air is drawn in from the sea as a **sea breeze**, which blows the flag.

Meanwhile, over the sea the air is still cool and sinking. This increases the air pressure over the sea. Up above, fed by the air rising over the land, high level winds blow out to sea to replace the sinking air.

People sitting on the beach will see clouds blowing out to sea at high levels, but they will feel the cool sea breeze blowing off the sea. They will see the flag flying in the opposite direction to the clouds. Winds are named with the direction *from* which they blow, so that is why the wind blowing the flag is called a sea breeze.

1 In Fig. 8.2 night has fallen.
a) Which will be warmer, land or sea?
b) Use the three points above to help you add arrows and labels to a copy of Fig. 8.2.
c) What name would you give to the breeze and why?

These breezes are local winds. They blow only up to 3 or 4 km inland or out to sea. Keep an eye out for them next time you are at the seaside.

Fig. 8.1 How a sea breeze is caused

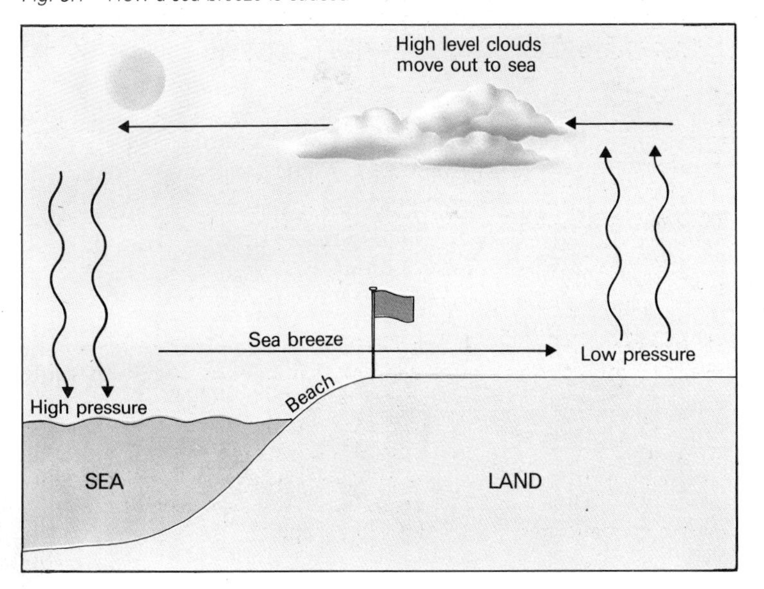

Fig. 8.2 Outline for Exercise 1

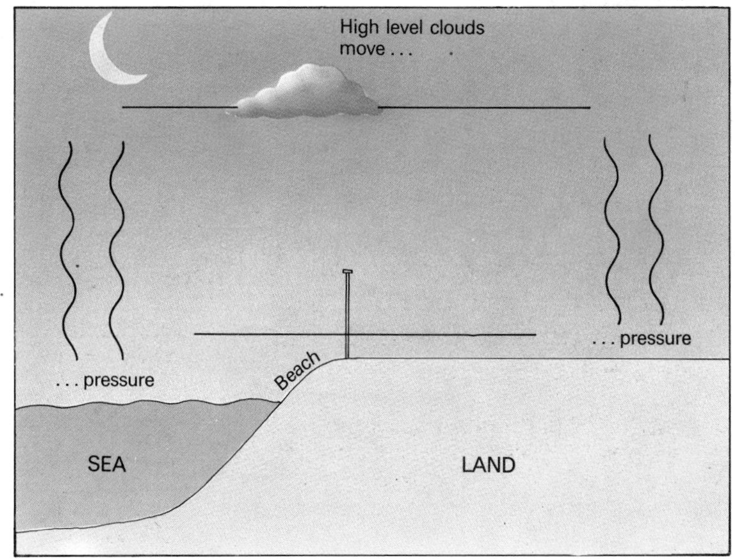

World-scale winds

Fig. 8.3 shows a section of the globe. Cold air sinks over the North pole, causing high pressure with winds blowing out from the polar ice. Hot air rises over the equator, causing low pressure. This draws in air from both sides of the equator to replace the rising air.

The two spirals or **cells** of revolving air are called the Polar Cell and the Hadley Cell. The latter is named after George Hadley who first had the idea that these cells existed.

2 Draw a diagram of a cross-section through the globe. Mark in the North and South poles and the equator. Add the pattern of winds and air currents you would expect to find.

Because the earth is spinning round, winds in the northern half of the world are diverted to their right. Fig. 8.4 shows that the cells, rather than running north–south, are tilted so that the winds blow from the north-east rather than from due north.

Hadley knew that in Britain the wind blows most frequently from the west and south-west (for example, the winds in the spit game, p. 74). He thought that another cell would be found between the Polar Cell and the Hadley Cell; this cell would rotate, powered by the cells on either side, like a cog-wheel between two other spinning cogs. As we shall see he was not quite right.

What really lies between the Polar Cell and the Hadley Cell? Part of the answer was found out during the Second World War when bombers flew at great height to be above the range of the simple radar of the time. What they found was often rather puzzling. For example, bombers from an airfield, like the one in Fig. 8.5, were ordered to take off, fly west for two hours at an air speed of 200 km per hour, and drop their bombs. At the height they were flying there were westerly winds blowing at the same speed.

3 a) On a copy of Fig. 8.5 mark a cross where you think their bombs fell.
b) When the bombers flew back at an air speed of 200 km per hour for two hours, where did they land? Mark the place on your copy.

Gradually weather men realized that they had discovered a westerly **jet stream** flowing at great speed from west to east about halfway between the equator and the North pole. It is about 10 km above the earth and is shown on Fig. 8.6 as the winding edge of a belt of westerly winds.

The position of jet streams is now included on most airline pilots' brief-

Fig. 8.3 Winds and air currents forming the Polar Cell and Hadley Cell

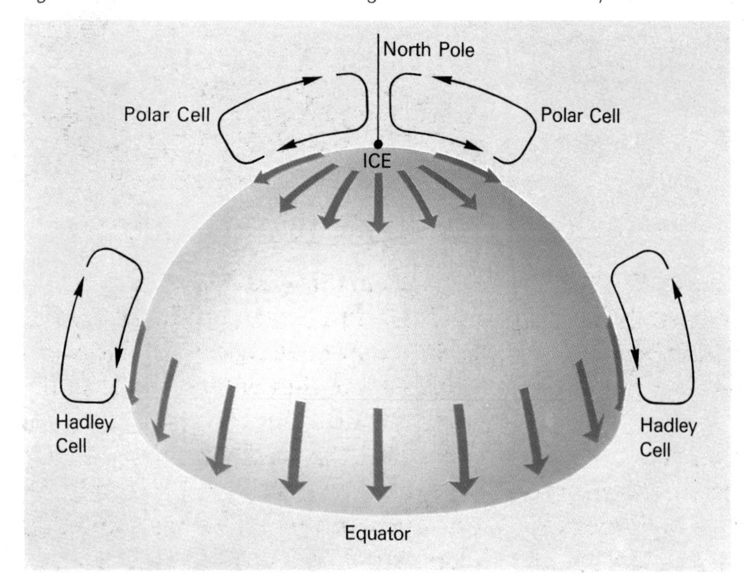

Fig. 8.4 The winds produced by the Polar Cell and Hadley Cell

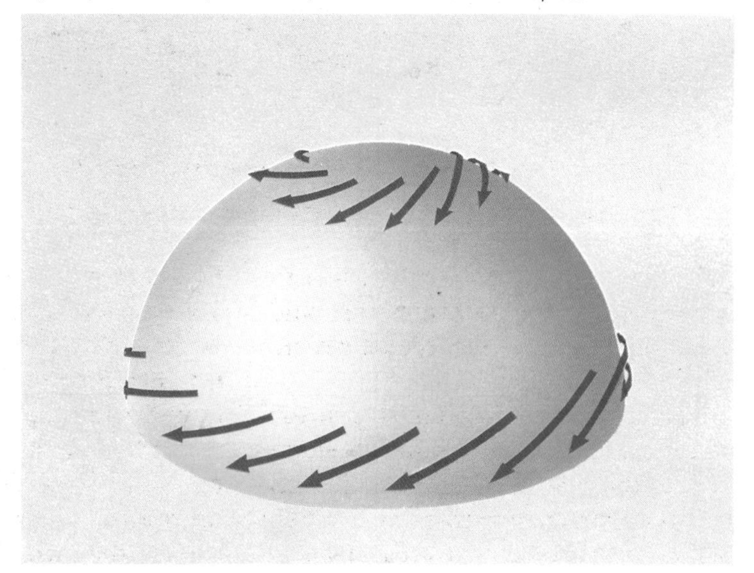

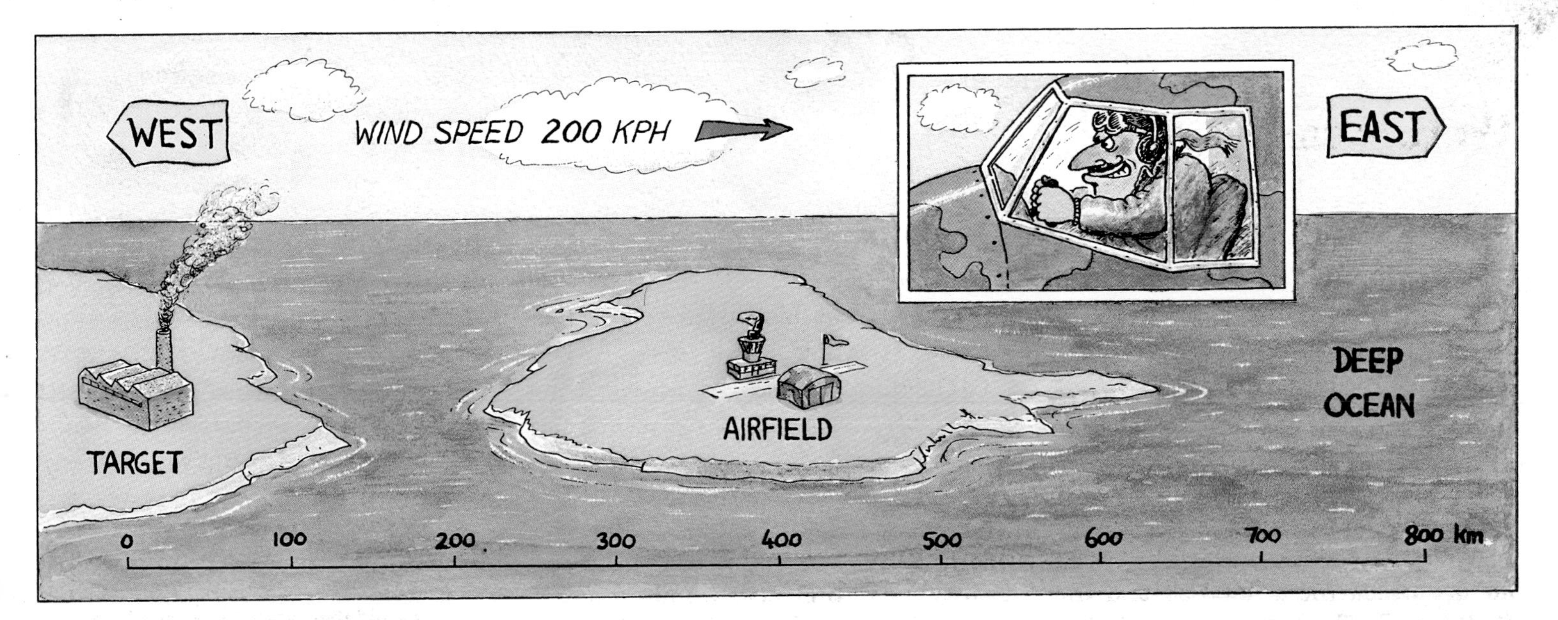

Fig. 8.5 Airfield and target for Exercise 3

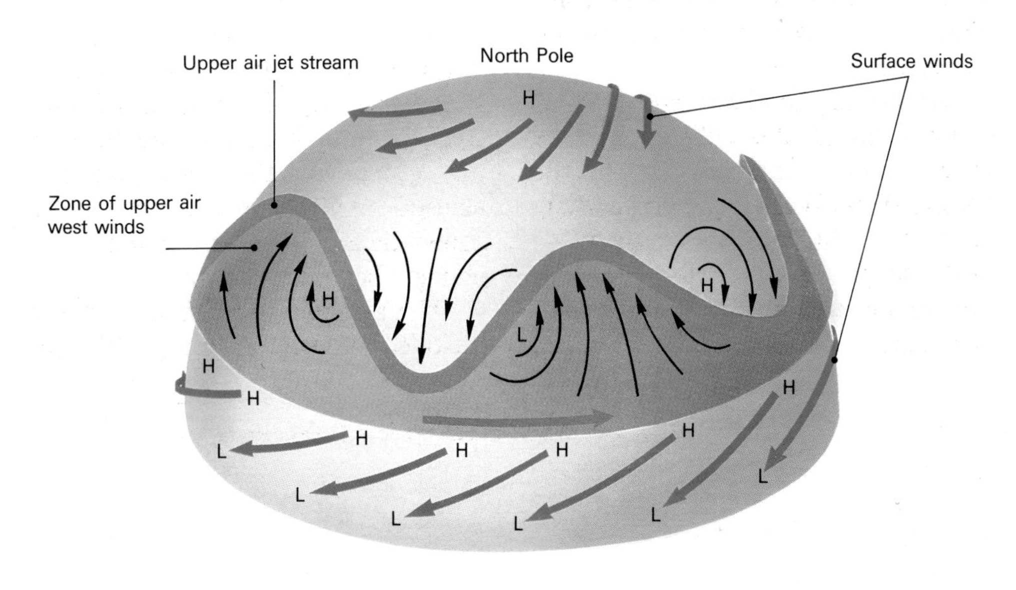

Fig. 8.6 How the jet stream forms areas of high and low pressure at the earth's surface

Fig. 8.7 High cirrus cloud is often a sign of bad weather

ings. Modern airliners use the winds to save fuel on their journey east and need to be able to avoid the jet stream when flying westwards.

The jet stream does not stay at the same height or the same position above the earth's surface. It rolls around in an endless corkscrew spiral. When it swings northwards, it tends to rise above the earth's surface causing low pressure to form below it; when it heads southwards, it tends to sink down and so high pressure areas form beneath it.

Although the jet stream is so far above us, it affects us all. As it pulls depressions and zones of high pressure round the world, it causes great changes in our weather.

High thin **cirrus clouds** (Chapter 4, p. 48) coming along ahead of a warm front are the sign of high winds in the upper air drawing the clouds out into long thin streaks. They are the visible sign that the jet stream is moving northwards and upwards, bringing along another depression (Fig. 8.7).

Pattern of winds and ocean currents

Fig. 8.9 shows the ocean currents which in many ways reflect the pattern of winds. It used to be thought that the winds blew the ocean currents along. As you will see, if you compare Figs. 8.8 and 8.9, there are many similarities.

More recently it has been realized that the ocean currents drive themselves by the cold water sinking and warm water rising, and then moving across the surface of the sea as warm currents. Indeed the ocean currents and water temperatures have far more influence on the air above than the winds have on the ocean.

Fig. 8.8 World pattern of prevailing winds

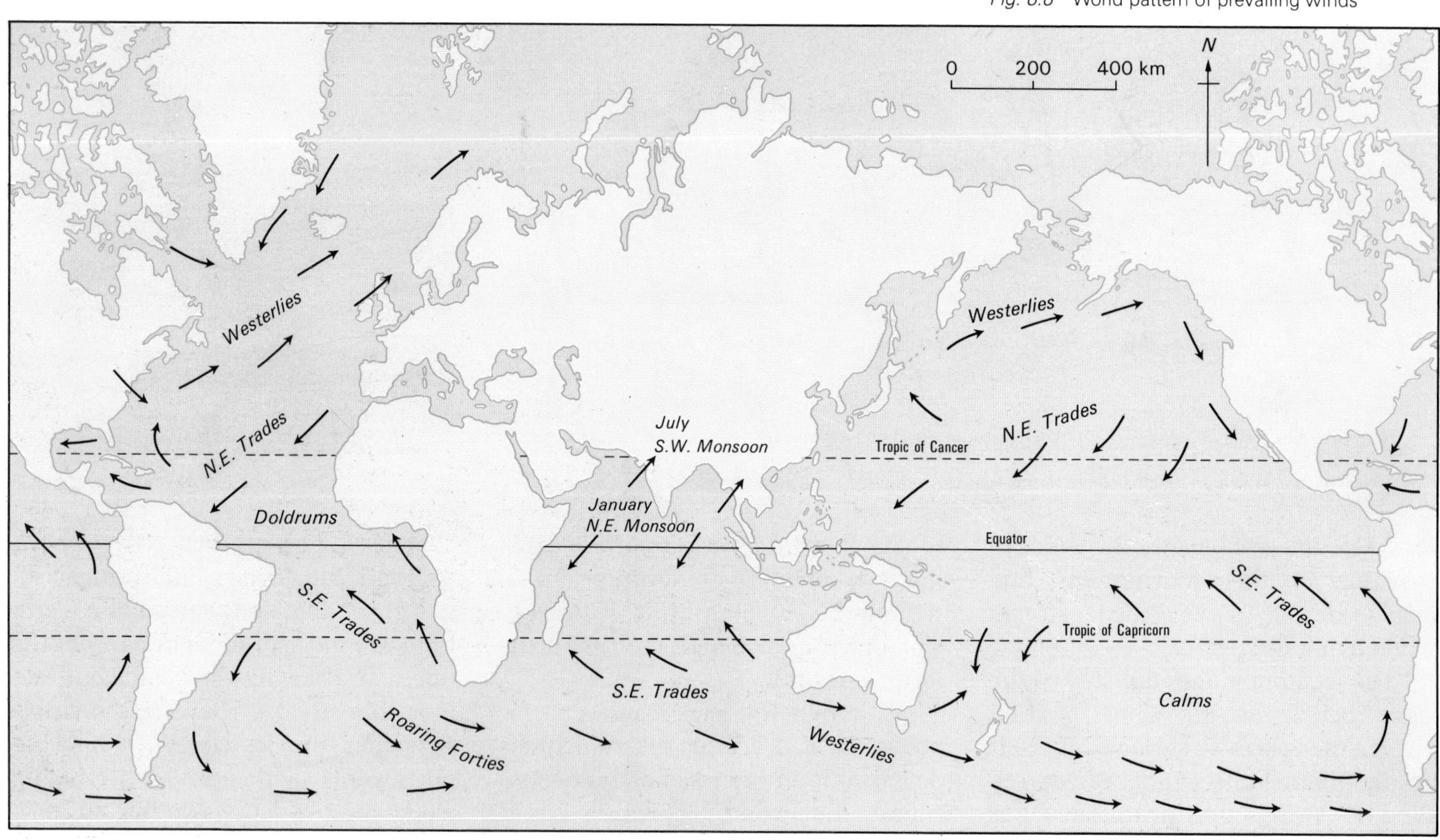

Round the world yacht race

The aim of the race is to go round the world in as few weeks as possible (Fig. 8.9).

1 You need a dice and a counter.
2 Start and finish at Britain.
3 Choose your intended route and throw the dice; each spot on the dice is one week's sailing.
4 Each time you throw the dice you must *start* off in the direction shown by the arrow in the square you occupy. After this you can carry on in any direction (vertically, diagonally or horizontally) from square to square, but not of course across the land.
5 To cross a square in the direction shown by the arrow is one week's sailing. To sail through in any other direction takes three weeks.
6 Squares labelled 'calms' require a six to cross or to move out from.
7 If you throw a one while in an area of tropical storms, you must spend that week in the same square while you ride out the storm.

4 As you play, keep a diary of the parts of the world you visit, the currents you use, and any storms that you meet. You should write about a line for each week that anything happens. At the end add up the number of weeks it has taken you to sail round the world.

Fig. 8.9 Base map for the round the world yacht race

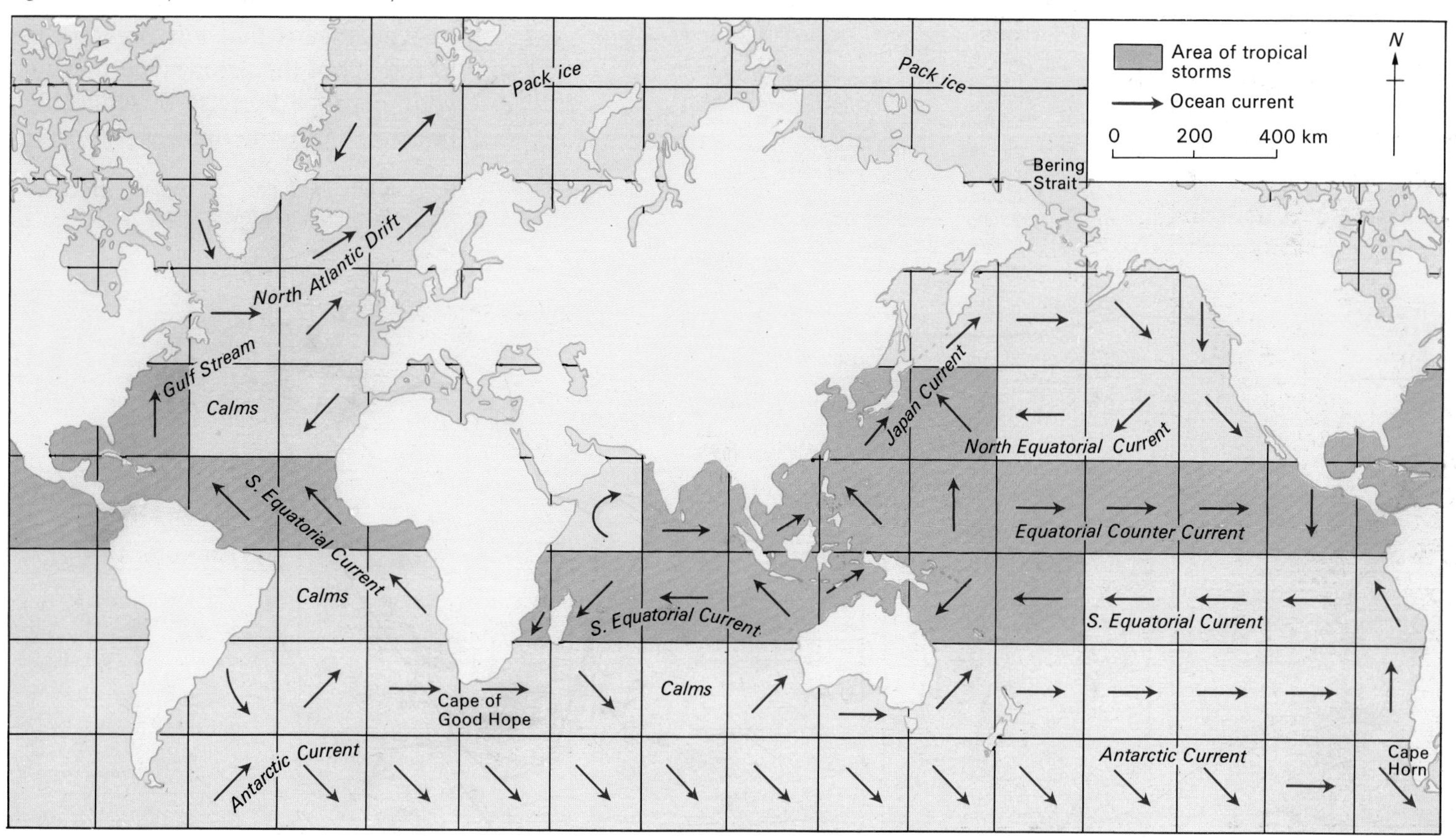

Wind speed

One way of estimating wind speed is to use the Beaufort wind scale (Fig. 8.11). It is worth making yourself completely familiar with the scale, since shipping forecasts and gale warnings mean so much more if you know it. Fig. 8.10 is an example of a shipping forecast.

5 a) Use Fig. 8.10 or listen to a shipping forecast on the radio. Make a note of the wind force and direction in each of the sea areas.
b) On a copy of Fig. 8.13 use arrows to show the wind directions and mark the force by the side of each arrow.

We have already seen that north of the equator the revolving earth deflects the wind to the right. A simple way to remember this is that when you stand with your back to the wind, the *low* pressure is on your *left*.

This means that low pressure lies to the left of the arrows you mark on your map for Exercise 5, while high pressure lies to the right.

6 Add the high and low pressure areas to your shipping forecast map.

Fig. 8.10 Shipping forecast, midnight 3/9/80

Viking	Becoming southerly force 4 to 5.
Forties	South-westerly force 3 or 4, increasing to 7.
Cromarty	North-westerly 4, backing to south-westerly 7.
Forth, Tyne and Dogger	South-westerly 4, increasing to force 5 and 6.
Fisher	Westerly 3 or 4.
Humber, Thames, Dover, Wight, Portland and Plymouth	West 4, increasing to 6.
Sole	South-westerly increasing to force 7, then veering west and decreasing to 4.
Lundy, Fastnet, Irish Sea	South-westerly 6, veering west later.
Shannon, Rockall	South-westerly 7, decreasing and veering west force 4 or 5.
Malin, Hebrides	Southerly 4, increasing 6 or 7 later.
Fair Isle	Westerly 4, backing south, increasing 6 later.
Faeroes	South-east 4, increasing to force 6 later.

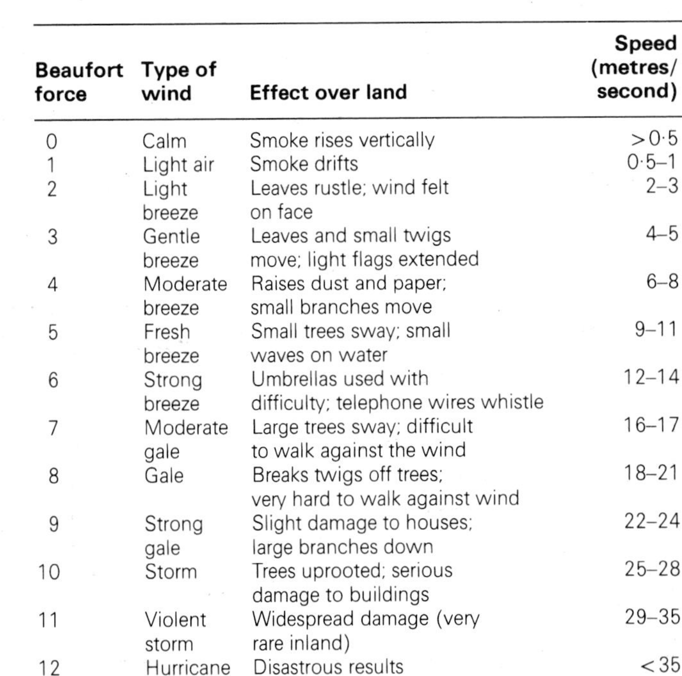

Beaufort force	Type of wind	Effect over land	Speed (metres/ second)
0	Calm	Smoke rises vertically	>0·5
1	Light air	Smoke drifts	0·5–1
2	Light breeze	Leaves rustle; wind felt on face	2–3
3	Gentle breeze	Leaves and small twigs move; light flags extended	4–5
4	Moderate breeze	Raises dust and paper; small branches move	6–8
5	Fresh breeze	Small trees sway; small waves on water	9–11
6	Strong breeze	Umbrellas used with difficulty; telephone wires whistle	12–14
7	Moderate gale	Large trees sway; difficult to walk against the wind	16–17
8	Gale	Breaks twigs off trees; very hard to walk against wind	18–21
9	Strong gale	Slight damage to houses; large branches down	22–24
10	Storm	Trees uprooted; serious damage to buildings	25–28
11	Violent storm	Widespread damage (very rare inland)	29–35
12	Hurricane	Disastrous results	<35

0: CALM.

1: LIGHT AIR.

Fig. 8.11 Beaufort scale: wind forces 0 to 12

The speed of the wind is measured on an **anemometer**, usually mounted on a tall pole or building.

Another type of anemometer, a hand-held one, is shown in Fig. 8.12. It is easy to carry and is very useful for fieldwork on local climate. For example, you can use it to measure the wind speed in different streets and round buildings.

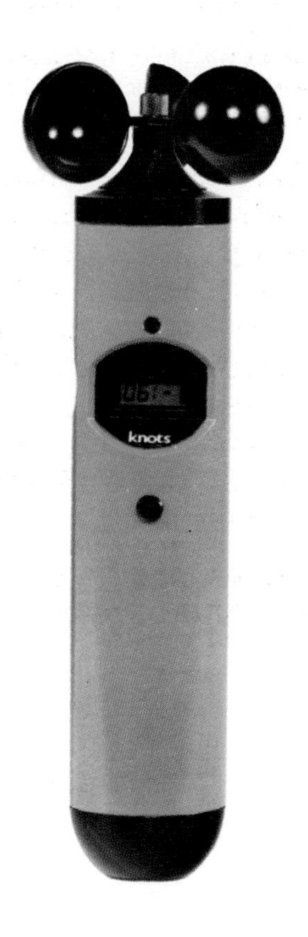

Fig. 8.12 A hand-held anemometer

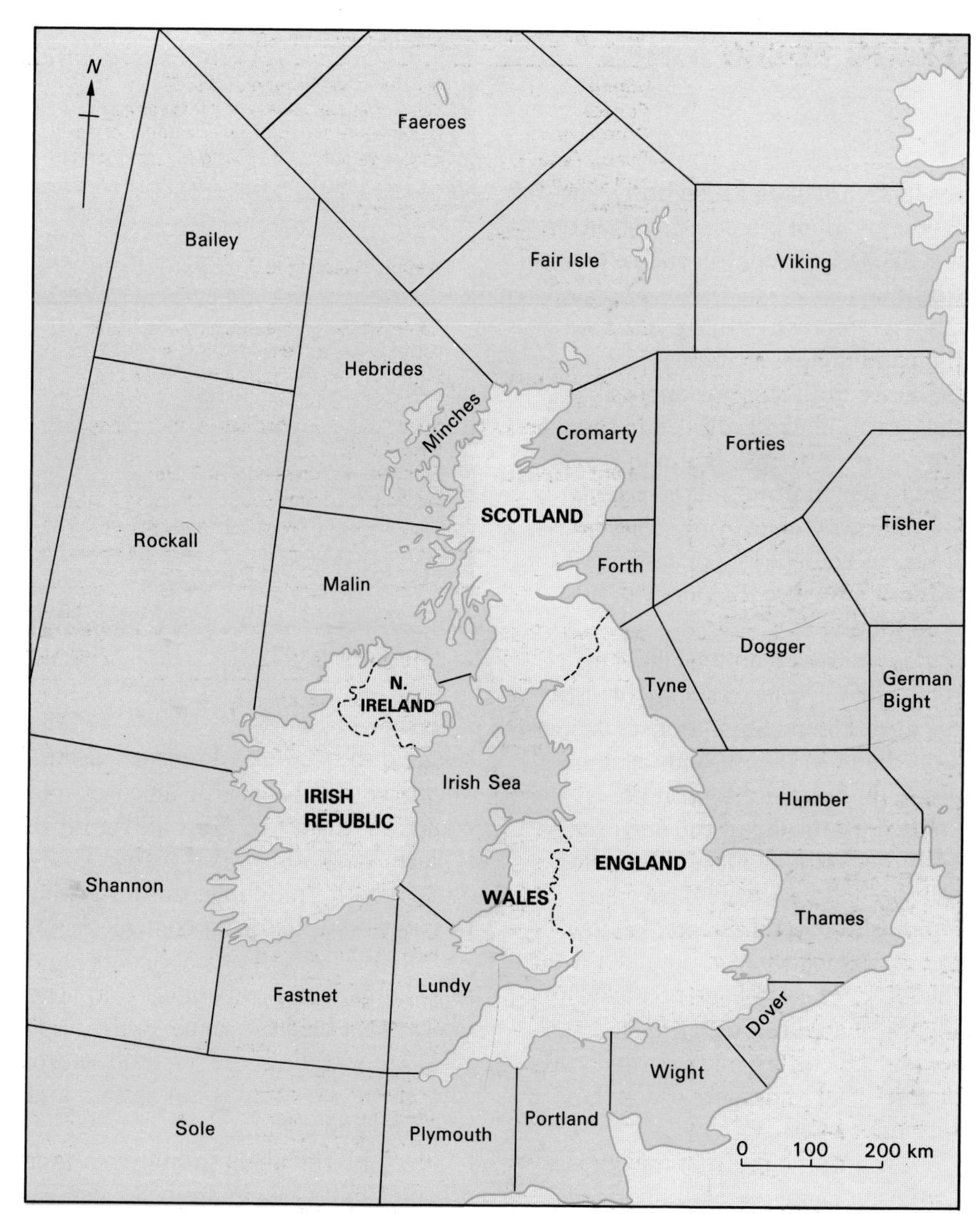

Fig. 8.13 Sea areas around Britain

Tropical cyclones

Cyclones are often called **hurricanes** or **typhoons**. Here is a description of what it is like to be caught in such a tropical storm.

'The impressive thing is the way the wind increases, increases, and increases until it seems impossible that it can blow harder—and then it does. One particular typhoon I was able to watch from the reasonably safe doorway of a strong stone house. It was as though I were looking at the stage of a theatre. I alone was motionless.

'Huge trees would be lifted up, roots and all, and go sailing off into the atmosphere. Small houses dashed after them. In the harbour, steamers of ten thousand tons, with noses pointed frantically at the open sea, with the anchors out and steaming full speed ahead, were piled stern foremost on top of the breakwater. Street cars left the tracks; down went the wires. Sheets of corrugated iron roofing soared through the air, sometimes cutting off the arms or legs of desperately hurrying pedestrians seeking shelter.

'Then in the midst of the wildest confusion came the ominous dead calm, which marked the so-called centre of the typhoon. But the quiet was only momentary. The other edge arrived, the wind reversed, and everything moveable went roaring back again.'

(Victor Heiser: *A Doctor's Odyssey*, Jonathan Cape)

In the Atlantic Ocean a cyclone can form in the following way. Warm sea near the Canary Islands warms the air above it. This warm air rises and cooler air rushes in from all round to replace it. Rather like water going down a bath's plughole, this air begins to spin round the rising centre.

As the air rises, it cools and releases its water vapour as rain. The energy that held the water in the air is released warming the air still more. The rising air carries on rising and spinning and cooling.

Because the air is spinning so fast, it begins to pull away from the centre, just like a conker spun round on a string tries to pull away from your fingers. In the same way the air trying to reach the centre of the spinning vortex can never reach it.

So the massive spiral, 80 km or so wide, spins across the warm tropical sea for weeks at a time. Only when it reaches dry land or very cold air will it lose its power.

Fig. 8.14 Satellite photograph of Hurricane Alicia over the Gulf of Mexico

7 a) Look at Fig. 8.14 and compare it with an atlas map of the same area.
b) Draw a sketch map to show the size of the cyclone shown in Fig. 8.14.

8 Fig. 8.9 (p. 89) shows the main areas of frequent cyclones as grey squares. On a world outline map mark the average number of cyclones that can be expected each year using the following information:

Area	Number
North-west Pacific	30
North-east Pacific	15
North-west Atlantic	9
South-west Indian Ocean	8
South-west Pacific	7
South-east Indian Ocean	6
Bay of Bengal	4
Arabian Sea	1

9 a) In the northern hemisphere there is a theory that water spins round the plug-hole in an anticlockwise direction. Next time you have a bath, check whether this is true.
b) If the whole class conducts a survey, does the water in more plugholes revolve anticlockwise or clockwise?
c) Which way is Hurricane Alicia rotating in Fig. 8.14?

Tornadoes

Although we seldom have hurricane force winds in Britain, we do have **tornadoes** (Fig. 8.15). For a tornado to occur there must be a violent uprush of air, often associated with thunderstorms. This can only be managed by air revolving in a fast corkscrew column like that in Fig. 8.16.

The width of a tornado's spiral is between half a metre and a few metres across, but within this the air may be spinning round at 300 km per hour. A tree caught in its path will have its top twisted completely off. Tornadoes move forwards at a rate of about 30 km an hour.

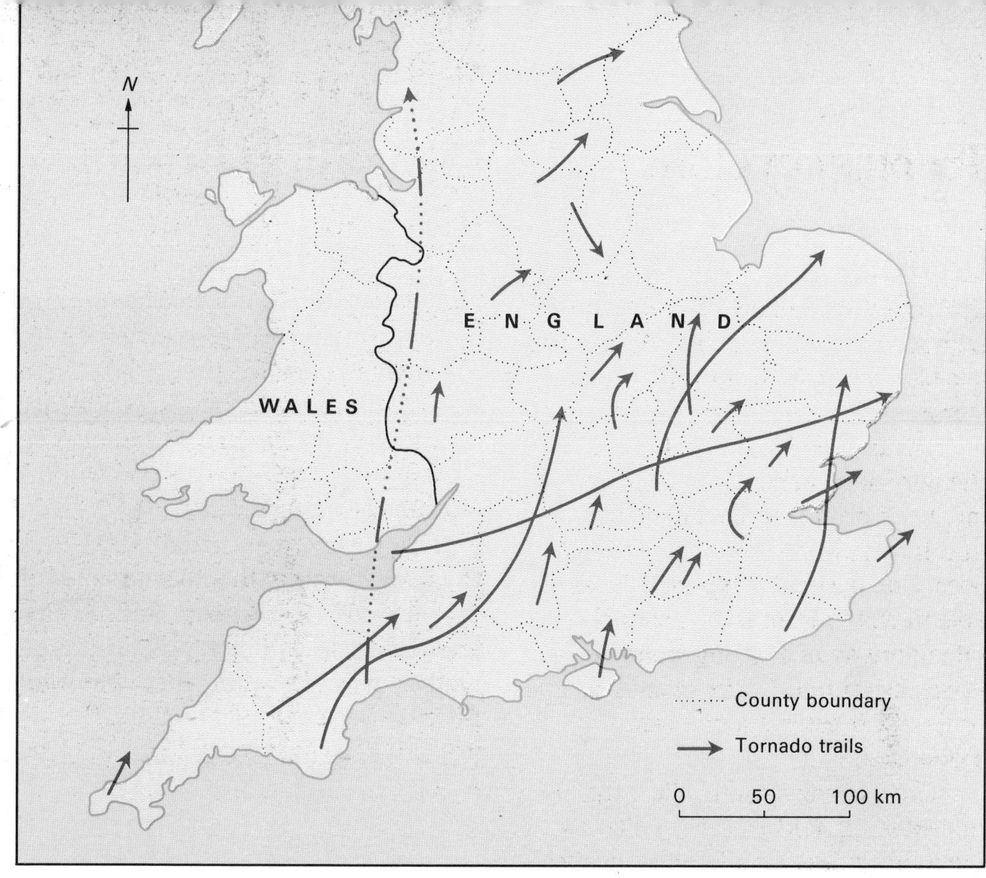

Fig. 8.15 Tracks of major tornadoes in England and Wales recorded in the last 400 years

Fig. 8.16 A small tornado or funnel cloud looming over Elgin, Scotland in May 1978

10 In which direction did most of the tornadoes travel in Fig. 8.15?

☆**11** Fig. 1.25 on p. 19 shows a funnel cloud on the right-hand side of the dust cloud erupting from Krakatoa. This is the first possible stage of a tornado. Why do you think this should have formed?

Summary exercises

12 Use the last chapter to help you to explain the meanings of the following words:

sea breeze	anemometer
land breeze	Beaufort scale
Hadley Cell	tropical cyclone
jet stream	tornado
ocean current	cirrus cloud

13 Make a tracing of the satellite photograph of the earth (Fig. 3.11, p. 39). On it label and mark any features you can identify as being mentioned in this chapter.

9
Drought

Drought in Britain

The poster in Fig. 9.1 was put up during a drought in Britain in 1976. Throughout most of the 1970s there was a drought in at least twenty countries in Africa. In 1982 a massive drought hit Australia. Is the world's climate becoming drier?

From May to August 1975 less than two-thirds of the normal amount of rain fell in Britain and people enjoyed their summer holidays. September was wet, but the winter of 1975/6 was very mild and dry.

The weather remained dry in 1976, apart from a wet May in the north of Britain. By June notices like that in Fig. 9.1 were appearing in the streets of London and the south of England. Hardly any grass grew in the the pastures, lawns and fields were burnt brown, and in forest and heathland the undergrowth was tinder dry.

The level of water in many reservoirs fell dangerously low (Fig. 9.2). On 6 August the Drought Act was passed allowing Water Boards to start

Fig. 9.2 A reservoir in the south of England during the drought

Fig. 9.1 Many of these posters were seen in 1976

Fig. 9.3 Standpipes were not so amusing to use once the novelty wore off

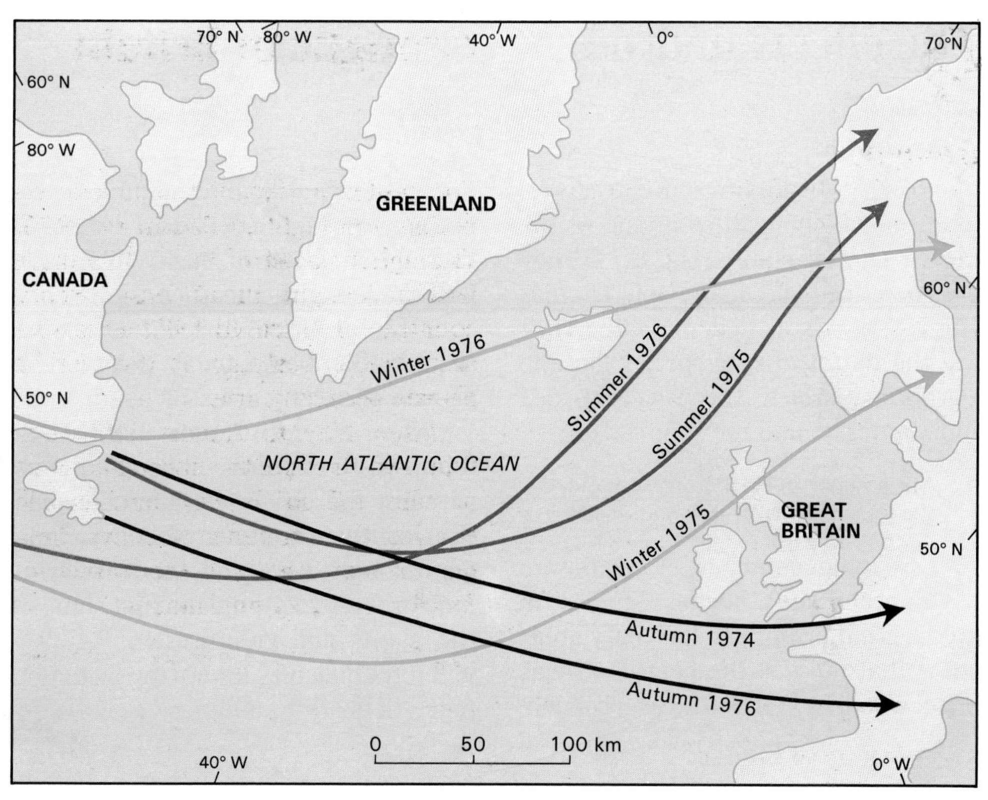

Fig. 9.4 Paths of the jet stream 1974-6

rationing water. Standpipes were set up in the streets and were the only source of water for most of the people of Devon for two weeks (Fig. 9.3).

By 29 August however the drought began to break and rain fell over much of the country. When the rain arrived, pastures which had been brown for months began to green up within hours. The effect was likened to a child's magic painting book with the instructions 'just apply water'. That autumn was very wet indeed and the level of soil moisture quickly returned to normal in most areas.

1 Use the text above to help you write a drought diary, summing up the main events during the course of the drought.

What had happened to cause the drought? A series of mild winters in Europe from 1971–5 had warmed the air more than usual. At the same time very cold weather in the North Pacific had cooled the water there and produced an area of very high pressure in that part of the world. This caused the westerly jet stream to change its course slightly.

As we saw in Chapter 8 (p. 87) the westerly jet stream guides the depressions which bring rain to Britain and north-west Europe. Fig. 9.4 shows that they moved much further to the north than usual, leaving Britain and Europe in a high pressure area with clear dry weather.

The hot summer weather warmed the sea by a full 2°C higher than average round the shores of Britain. This increased the moisture in the air and produced the very heavy autumn rains after the drought.

2 Use Fig. 9.4 to help you describe the course of the jet stream in:
a) summer 1975.
b) winter 1975.
c) summer 1976.
d) autumn 1976.
e) winter 1976.

As a general rule when the jet stream is aloft over Britain, the air below it is warm and wet. Lack of a jet stream gives dry clear weather; cold in winter, hot in summer.

3 Use the paragraph above and Fig. 9.4 to help you describe the weather for the periods listed in Exercise 2.

The effects of drought on heathland

The Dorset heathlands are tracts of land on which heather is one of the most common plants (Fig. 9.7). They are described in many of Thomas Hardy's novels as Egdon Heath. They are the haunt of rare species like the Dartford Warbler, the Silver-studded Blue butterfly, and the sand lizard.

4 Make a copy of Fig. 9.5 showing only the Dorset animals, birds, and insects.

Low-lying stretches of heath are often peaty and boggy. Studies of pollen grain trapped in the peaty areas show that most of the heathland was once woodland which was probably destroyed by fire. Indeed, unless the heaths are regularly burnt by small fires, birch saplings begin to invade and shade out the heather.

Fig. 9.6 shows the extent of the heathland in Dorset today. A long time ago all these stretches of heath were joined together. Animals could roam and seeds could drift or be carried all through the area. The gaps between the heathlands occurred when land was ploughed up and used as farmland or forest. This means that some of the plants and creatures in Fig. 9.5 will have difficulty in moving from one patch of heath to another.

5 Which of the plants and creatures shown on Fig. 9.5 would find it easy to colonize from one patch of heathland to another? Give reasons for your choice.

Every day during the summer several thousand people find the heaths a pleasant place for a picnic and accidental fires are quite common. In the drought of 1976 much of the heather died because of lack of moisture and the heathlands became tinder dry. The fires that form a part of most summers were more frequent and violent and whole patches of heathland were destroyed.

One of these was Middlebere Heath

Fig. 9.5 A heathland food web

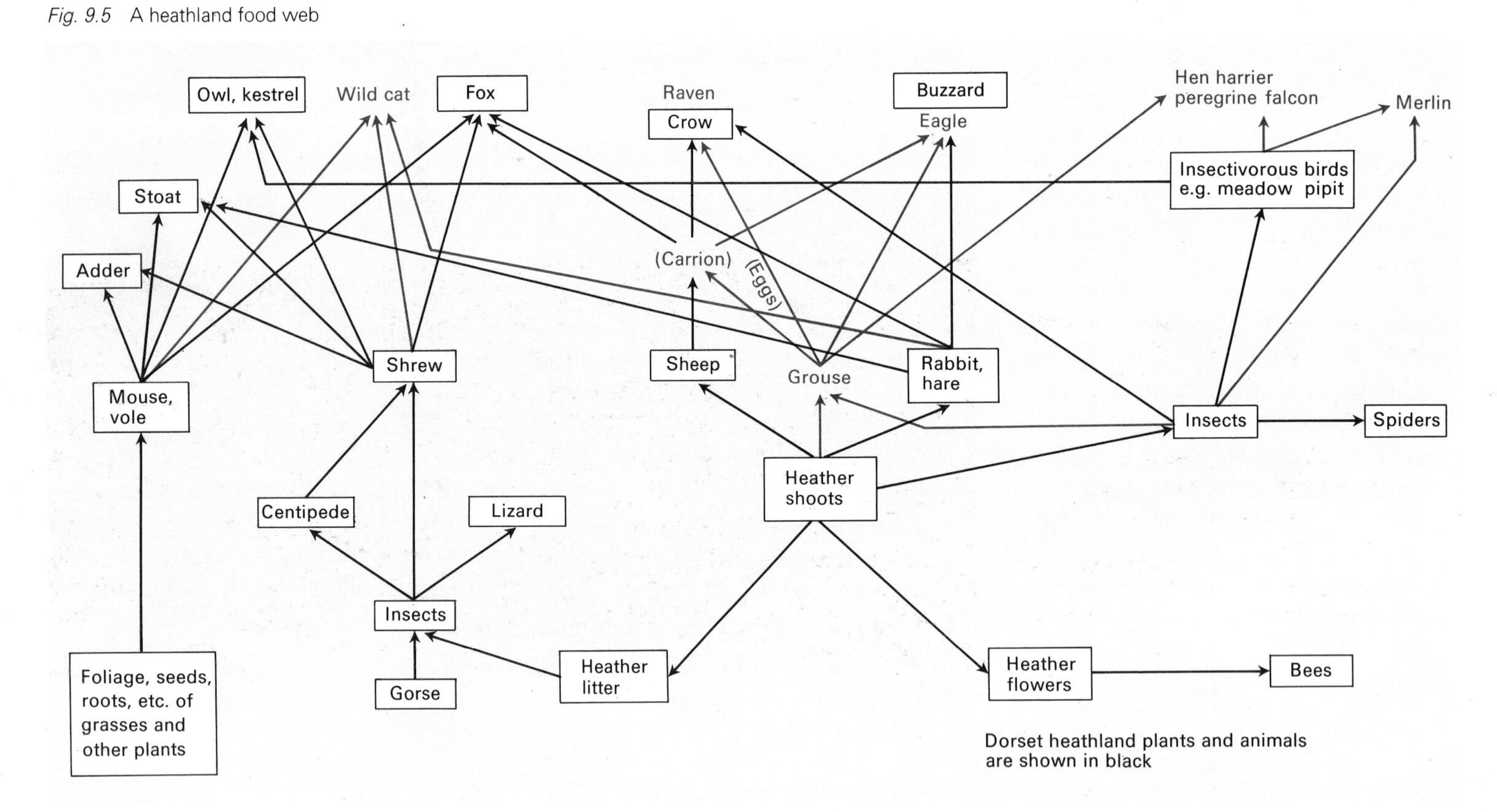

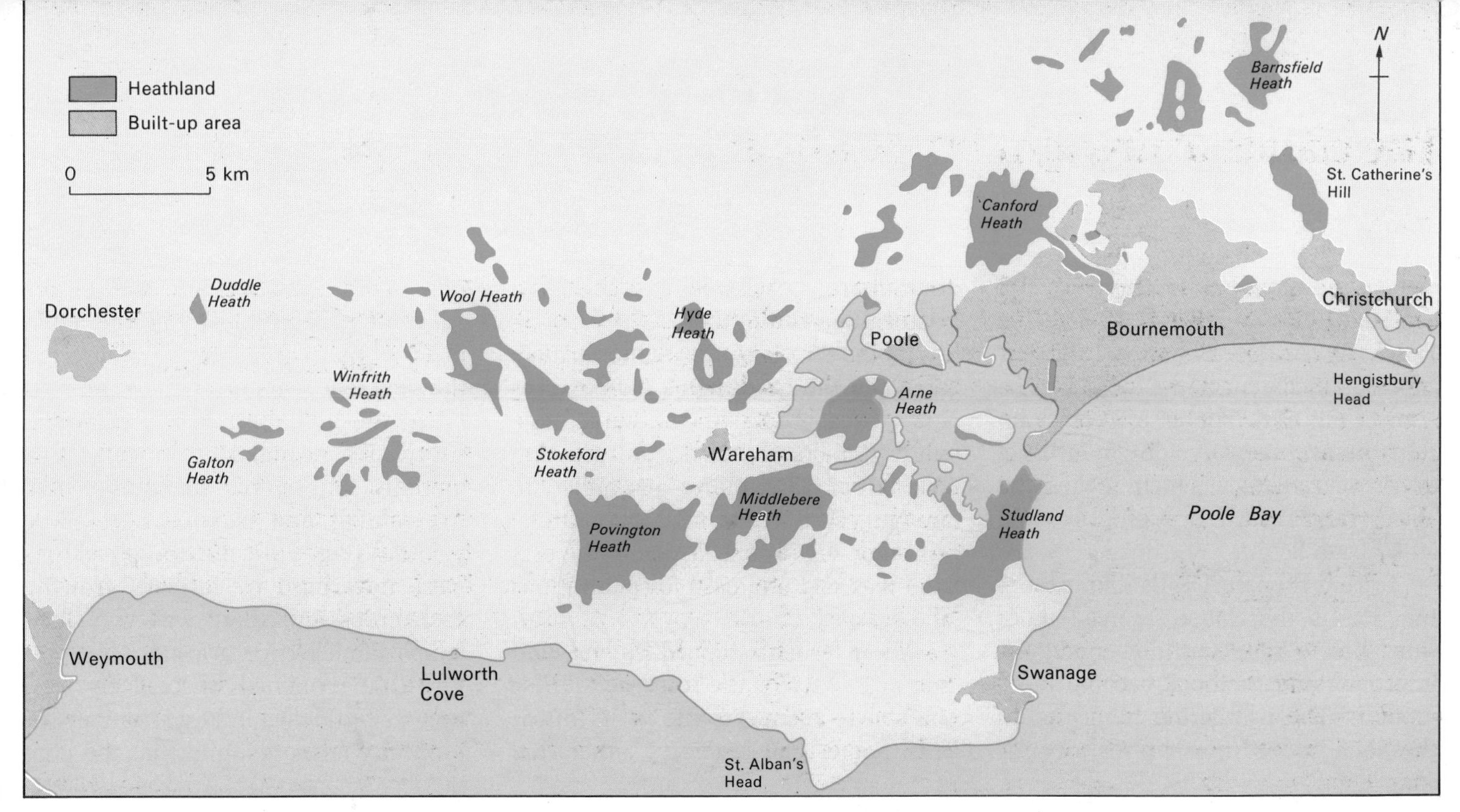

Fig. 9.6 The heathlands of Dorset

(Fig. 9.6). The peaty soils were burnt to great depths and the roots and seeds of the heather were utterly destroyed. As a result of being so far from other patches of heathland the heather has found it difficult to recolonize this heath.

6 a) In what ways have people contributed to the destruction of Middlebere Heath?
b) How many of these actions were done on purpose and how many by accident?

7 a) How would Middlebere Heath have recovered from such a devastating fire before man created the pattern of heaths shown in Fig. 9.6?
b) If it were not for the gaps between the heathlands made by man, how could the fire on Middlebere Heath have been even more disastrous?

Fig. 9.7 Late summer heather brightens Studland Heath in Dorset

The Sahel drought

Stretching along the southern edge of the Sahara desert is a zone called the Sahel, an Arabic word for 'edge' or 'shore'. The extent of the Sahel can be seen in Fig. 9.8. The area receives no rain on average for eight months of every year, a fact which makes the 'disastrous' British drought of 1976 look insignificant.

During the late 1950s and most of the 1960s the Sahel received more rain than usual. Farming conditions improved and methods became more modern. The wandering herdsmen of the Sahel moved northwards towards the Sahara to take advantage of the pastures appearing there.

In the south of the Sahel farmers ploughed the herdsmen's old pastures and sowed crops. At the same time many farmers changed from growing food crops like maize or millet to growing cash crops like groundnuts, cotton or sugar cane. Much of the best land was sold and used for plantations of these cash crops.

Many farmers stopped leaving land fallow for their cattle to graze and for the soil to recover some of its nutrients. Some used fertilizers, others did not, and the soil became exhausted. All went well until in 1968 and again in 1969 the rainfall was low. Through the next few years drought set in.

In the past the cattle herders had survived a drought by moving their animals southwards to areas where the rainfall and pasture were more plentiful (Fig. 9.8). But these pastures were now used by farmers growing cash crops and there was very little fallow land left for grazing. Any pasture that remained was quickly used up and cattle died in large numbers.

Many farmers found that the poor

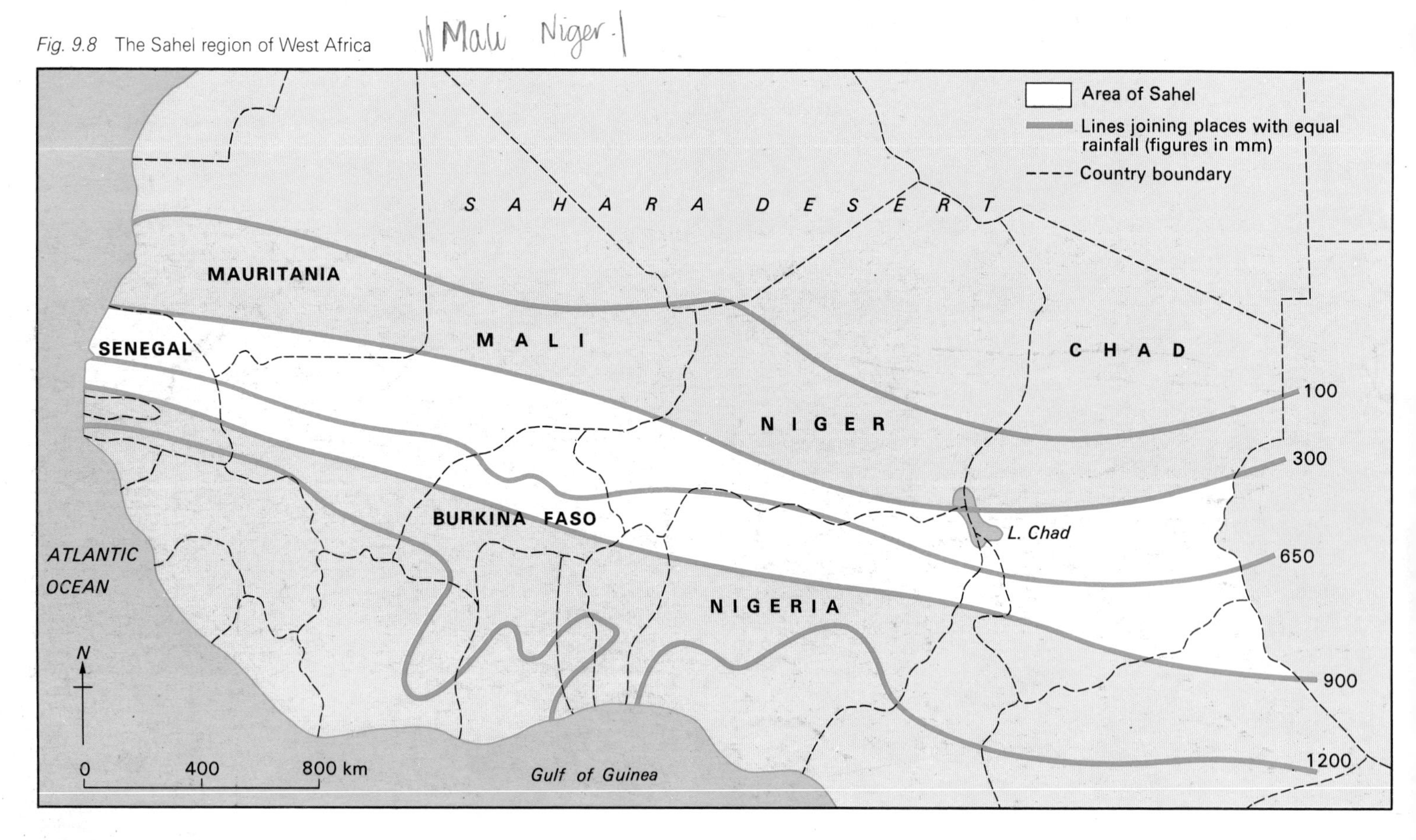

Fig. 9.8 The Sahel region of West Africa

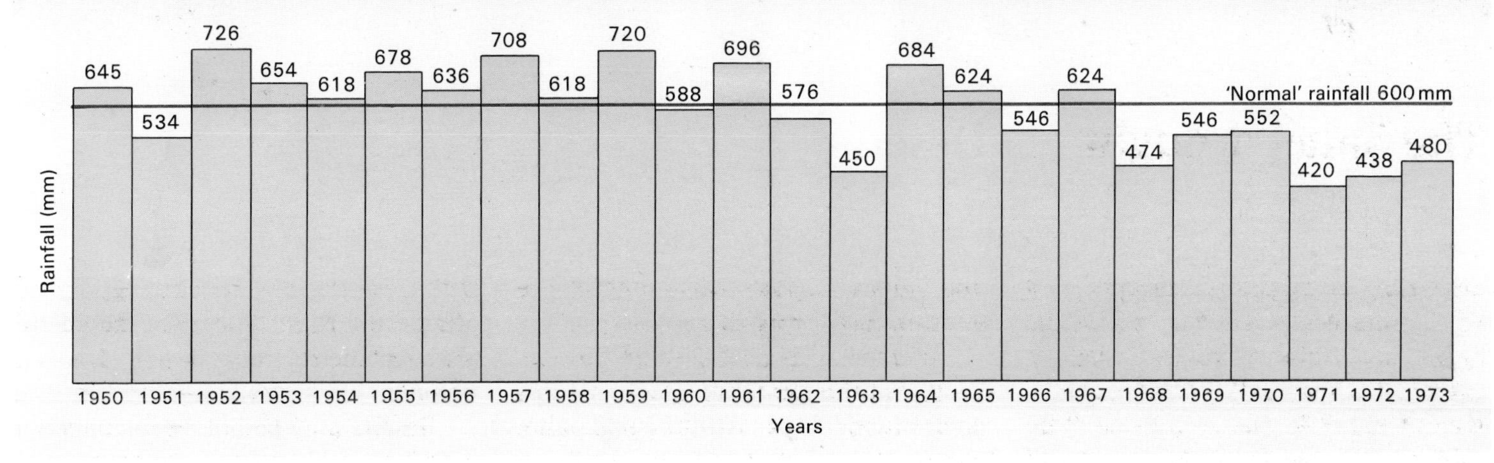

Fig. 9.9 Rainfall in the Sahel from 1950 to 1973

soil they were left with would grow plants no longer; the soil became bare and erosion set in. Cash crops like cotton could not be eaten; often they failed completely and so could not even be sold in exchange for food. The rainfall pattern shown in Fig. 9.9 resulted in famine in all the countries of the Sahel. There was famine on a large scale.

8 Make a table with two columns, the first showing changes in agriculture carried out in the late 1950s and the 1960s, the second showing the disasters that occurred in the drought years of the 1970s as a result of each of the changes.

The rainfall figures in Fig. 9.9 show the rainfall decreasing over the years. Some people think that this is the first sign of a change in the climate as the Sahara desert moves southward.

Other people say that droughts like that in the 1970s have occurred quite regularly in this area. Rainfall figures show a period of drought in the years between 1910 and 1915 and another during the 1940s.

What has changed is not the climate but the farming methods. Traditional methods meant that the herdsmen and farmers survived such droughts in the past.

Fig. 9.10 Improvised water supply during the Sahel drought

Fig. 9.11 The drought brought death to thousands of animals

Deserts

Fig. 9.12 Massive sand dunes in the Namib Desert

9 Write down ten things that come into your head when you think of a desert.

We normally think of deserts as areas of sand dunes with the odd palm tree and camel trains leaving trails of footprints through the endless wastes. Sand desert, known as **erg** to Bedouin Arabs, consists of huge seas of sand with dune after dune (Fig. 9.12).

Most of the deserts shown on Fig. 9.13 are not sand desert and never have been. Enormous areas of desert are floored with gravel (**reg**), while others are bare rock for kilometre after kilometre; rock desert is called **hamada**.

Deserts may be hot or cold. Even a hot desert like the Sahara, despite having days of searing, blinding heat, has nights when the cold is numbing and the wind is often icy.

Deserts are sometimes described as being areas where less than 25 cm of rain falls in an average year. A better definition is to say that they are areas where the amount of water that could

Fig. 9.13 (below) World pattern of deserts

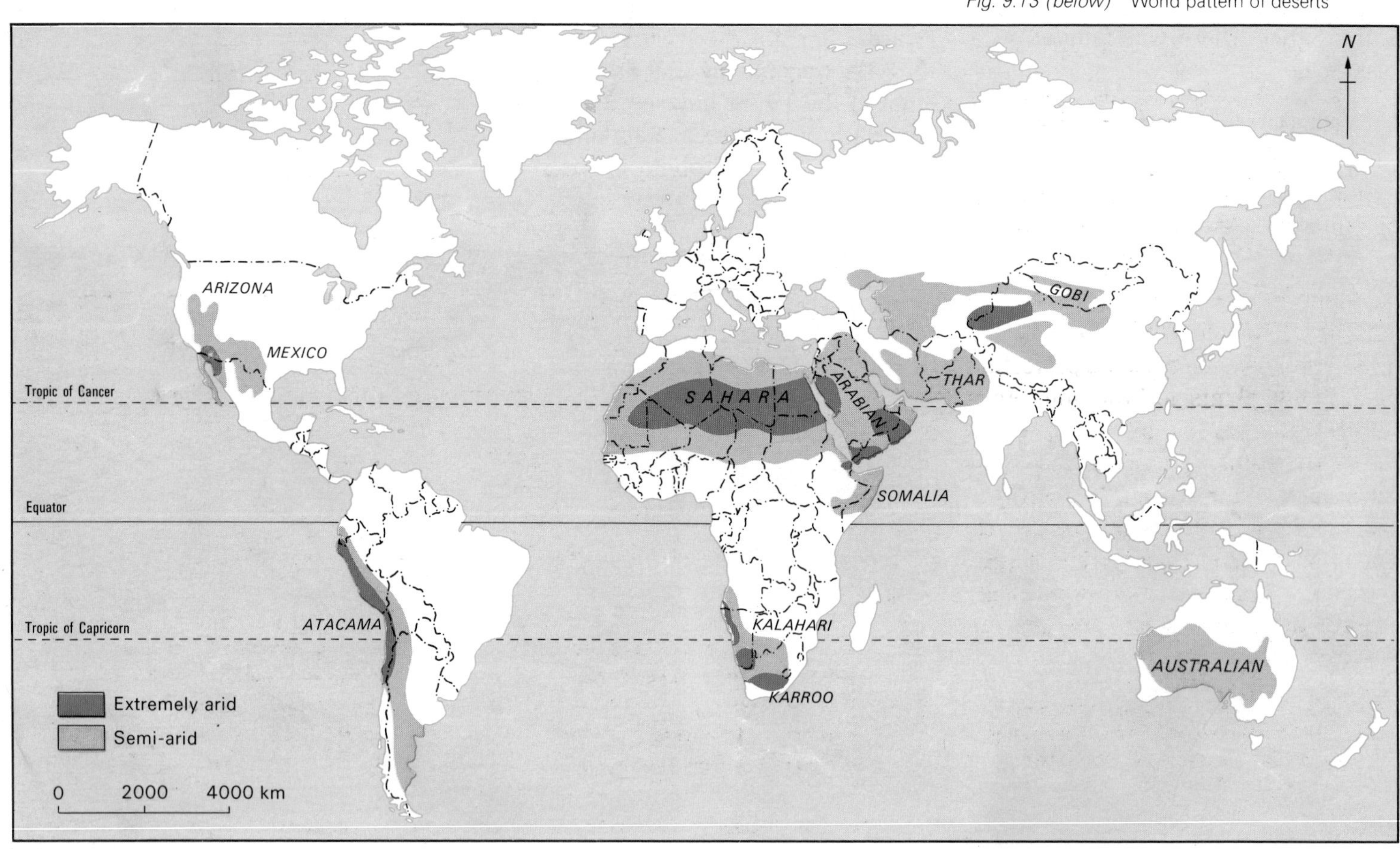

evaporate is greater than the amount of rain that falls.

There are three main reasons for an area being desert:

i) Areas with high pressure, such as the Sahara Desert, have winds blowing out from land to sea. This means that moist air has no chance to blow in from the sea and bring rain.

ii) Mountains may form a barrier preventing winds from carrying rain over them to the land in the **rain shadow** area behind. The Patagonian desert in South America is an example of a rain shadow desert.

iii) Some winds blowing in towards the land pass over cold currents in the ocean. The wind is thus cooled and is not able to carry much moisture. The Atacama desert is a good example of the type of desert that results.

10 a) With the help of an atlas decide why there are deserts in each of the areas shown on Fig. 9.13.
b) Sum up your ideas in a table with two columns, one giving the name and location of the desert and the other giving its possible causes.

Man-made deserts also exist. By cutting down trees and ploughing up grassland, man removes the protection of growing plants for the soil. As a result any rain runs over the soil rather than sinking down into it. This running water may wash away the soil or it may form rills or deep gullies. These drain water from the soil, which dries out and blows away more easily. The removal of the soil means that no plants can grow.

11 The boxes in the flow diagram (Fig. 9.14) have been left blank. On a copy of Fig. 9.14 write the correct labels in the boxes.

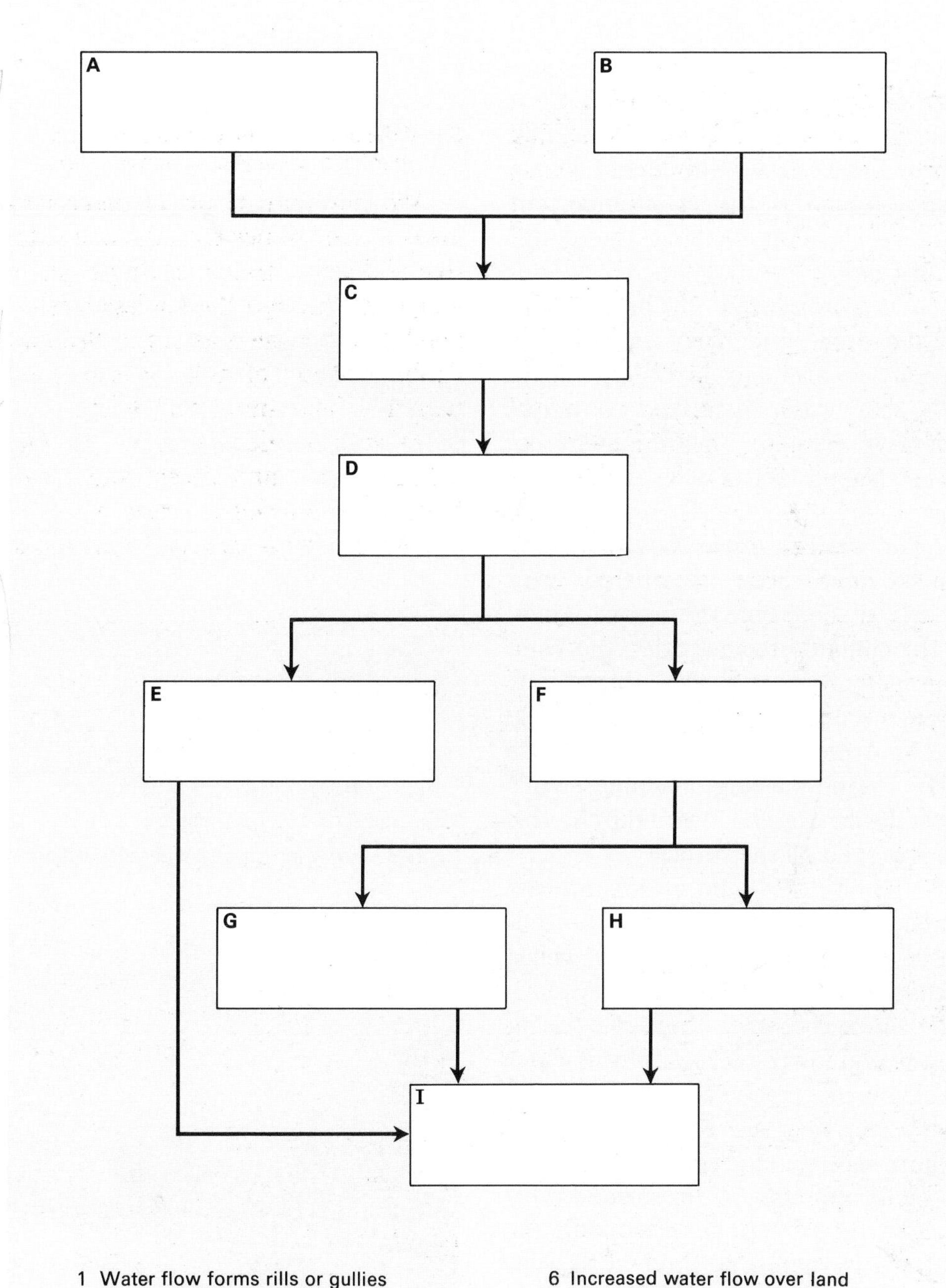

Fig. 9.14 Flow diagram (for Exercise 11) to show how people can turn places into deserts

Erosion in deserts

The erosion described in Fig. 9.14 is caused by running water and wind. There has been a great deal of debate about which is the most important form of erosion shaping the desert landscape.

Wind, as we saw in Chapter 5 (p. 62), can lift and carry sand grains away. We also saw in Chapter 6 (p. 77) that water is a very powerful agent of erosion. The argument between supporters of wind and water has swung to and fro like a pendulum.

For wind to erode rock, the rocks must have been weathered very deeply at some time in the past. Then, if the climate becomes drier, the wind can blow much of the weathered rock away, scouring out a hollow (Fig. 9.15). Armed with sand, the wind acts like a sand-blasting machine, while dust devils (Fig. 9.17) and dust storms lift the sand off the surface.

The lifting of sand grains into the air is called **deflation**, while the hollows which are formed are called deflation hollows (Fig. 9.15). Gradually these become deeper, exposing lower and lower rocks until the water table may be reached.

When they are blown across the plains, most sand grains are lifted only a few centimetres off the ground. This means they often bounce along the surface by a process known as **saltation**. The impact of one sand grain hitting the ground may well cause another sand grain to bounce up and be carried on by the wind.

Wind action armed with sand can result in unusual rock formations. For

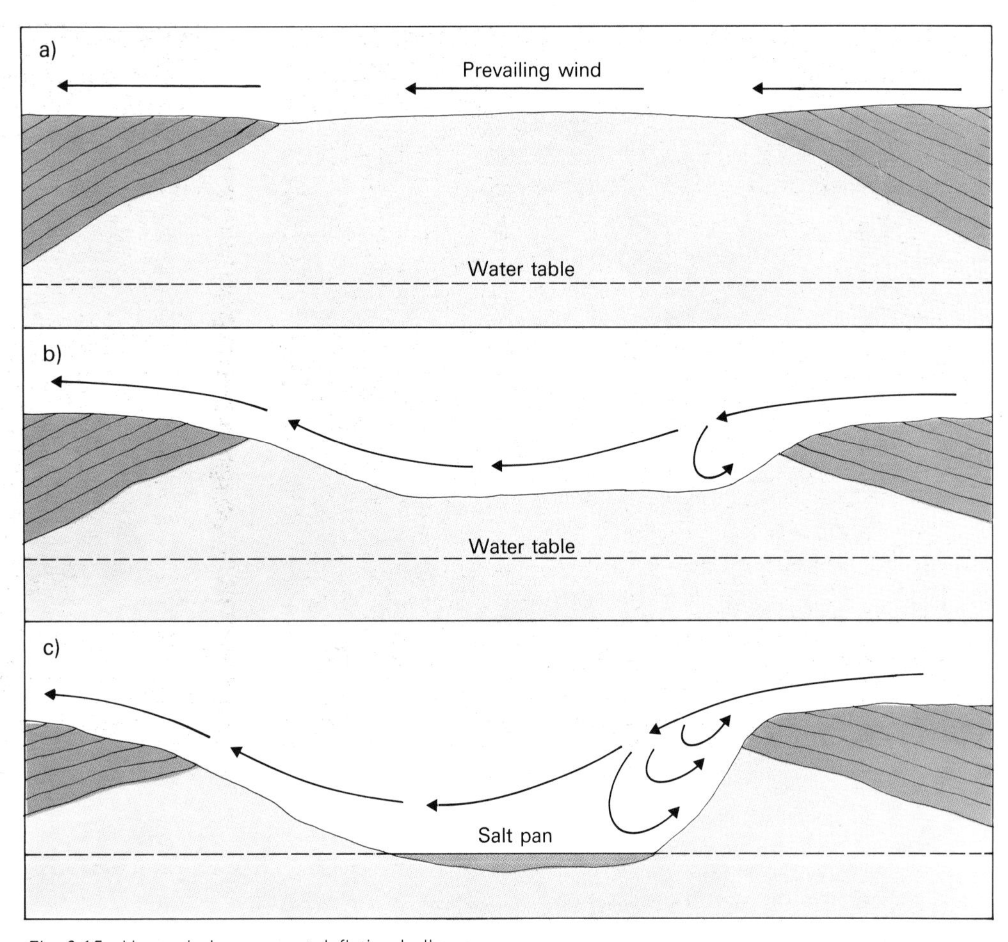

Fig. 9.15 How wind scours out deflation hollows

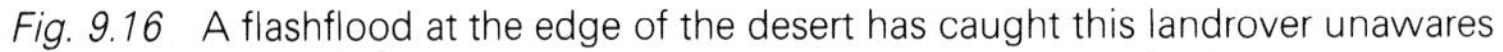

Fig. 9.16 A flashflood at the edge of the desert has caught this landrover unawares

Fig. 9.17 A dust devil in the Amboseli National Park, Kenya

Fig. 9.18 A pedestal rock in the Arizona Desert, USA

example, harder or more resistant rocks may form mushroom or **pedestal rocks** like that in Fig. 9.18.

Supporters of the importance of water erosion in deserts say that many so-called 'deflation hollows' could have been formed by earth movements, by limestone being dissolved, or by other means without deflation having occurred at all. Most wind action, they claim, takes place on a very small scale, adding detail to a landscape that has already been shaped.

For example, they say that dry valleys, known in deserts as **wadis**, must have been formed by water. Deserts may have a low annual rainfall, but when the rain does come, the whole amount may fall in one storm. Then sudden torrents of water or flash floods (Fig. 9.16) sweep down the wadis.

The hill slope that we saw in Fig. 6.11 (p. 70) is also a very common form in deserts. Water may move in sheet floods down the constant slope, carrying weathered rock material from the free face to the foot slope. The foot slope is called the **bahada** in deserts.

Recently the pendulum has swung back more in favour of wind erosion. Satellite photographs and aerial photographs show large-scale wind erosion with vast areas moulded by the prevailing winds and huge plumes of windblown dust stretching out across the Atlantic from the western Sahara. Many of the sediments drilled from the depths of the ocean reveal that they were once wind-blown dust.

Violent rainstorms do occur in desert areas but mainly near the edges. It may be that the water-formed features of deserts date from the wetter times of the past. Now they form a fossil landscape preserved by the dryness of the desert.

12 On a copy of Fig. 2.25 (p. 32) sum up the arguments of the wind and water supporters.

Drought in Australia

The wheat farmers of South Australia spend a great deal on irrigation and water conservation. Some wheat farmers manage to make a good living in spite of having droughts in four years out of ten.

If rain has not come by June they realize they are going to have a bad year. Their response to a drought is to live off the previous year's profits and off borrowed money if their own runs out. As soon as any rain appears, they put in a crop and hope for the best.

Such a rapid response to the end of a drought can have unhappy results, as it did for sheep farmers in eastern Australia in the 1940s.

'Comes a drought, small creeks and waterholes dry up and the animals surround those that remain. Soon grass and shrubs are eaten up, and around the watering place is a bare circle of baked earth. At its circumference there is herbage and the animals come in to drink and then walk back to the feeding ground.

'The bare circle is widening. Soon it extends in a radius of three or four miles from the water and the animals begin to fear the journey. Until driven by thirst, they do not leave the herbage, and until driven by hunger, they do not leave the watering place.

'Soon they are dying by thousands and hundreds of thousands. Had they been less numerous the radius of the bare circle would have been smaller and they would have survived the drought.

'Comes the rain and in a week or so the bare earth is once more green. Seeds in the ground have sprouted. More stock is bought and the young shoots are eaten up before the herbage has seeded.

'These droughts are recurrent, and so also became the practice of restocking until, in some areas, there was no new growth after the rain because there were no more seeds in the earth. So the baked soil crumbled and was blown away. In place of the station or sheep run there is a desert.

In recent years the lights of Adelaide have had to be turned on in the middle of the day, so dark the sky when the soil of Australia is passing out to sea.'

(Harry Sutherland: *Southward Journey*)

Fig. 9.19 In 1983, after five years of the worst drought in living memory, Australian farmers had to shoot thousands of stock

13 a) Sum up how drought and overstocking have led to soil erosion; use a flow diagram to illustrate the chain of events.
b) In what ways could this chain be broken and disaster prevented?
c) How would the grass have recovered if man had not interfered?

Summary exercises

14 Use the last chapter to help you to explain the meanings of the following words:

drought	sand dune
heathland	erg
soil erosion	reg
Sahel	deflation
desert	saltation

15 *'In some cases, the aid itself has caused new problems ... boreholes tend to bring large numbers of animals from different herds together which makes it easy for disease to spread from herd to herd, and pasture around the holes tends to become overgrazed. This leads to erosion; without the roots of plants to bind the fragile soil together, soil is washed away during the short rainy season reducing the potential for producing new grazing the following year.'*
(*The Mali Cattle Game*, Christian Aid pamphlet, 1983)
a) Produce a flow diagram to show how these boreholes in the Sahel, which are meant to improve pasture, can eventually lead to problems.
b) Which other section of this chapter has illustrated similar problems?

10
Floods

Floods in the news

Stories about droughts in Britain are far less often in the news than stories about floods. Hardly a year goes by without a flood becoming headline news (Fig. 10.1). Floods can cause great misery and hardship. We need to study them to see how they can be stopped from doing so much damage.

1 a) Use your atlas to pinpoint on the map the places described in the newspaper article.
b) Make a list of the types of damage caused by the storm.

Fig. 10.1 A report in the *Daily Express*, 28/12/79

Trapped couple drown in kitchen as storm trail of havoc leaves thousands homeless

FLOOD WALL OF DEATH

By Ashley Walton

A fearsome storm hit southern Britain yesterday—24 hours of death and destruction by wind and rain.

South Wales caught the full force of it, and last night police described the area as 'one big swamp' with thousands homeless.

Elderly widow Mrs Gladys Jones and lodger Mr Danny Jones were trapped in the kitchen and drowned when a wall of water from a bursting river engulfed their cottage at Rhydycar near Merthyr.

'They did not have a chance,' a fireman said: 'We had to wade up to our necks in the house to recover their bodies.'

As the village was evacuated three-month-old Rhiannon Nicholas, floating in a pram in her bedroom, was rescued through the roof.

In Monmouth a woman was missing after her car plunged into the River Usk.

At Trehafod 900 people had to leave their homes.

Aberfan, where 144 people died in an avalanche of coal slurry in 1966, was flooded, and hundreds were evacuated.

At Brecon on the Usk RAF helicopters lifted out troops from a barracks; two girls trapped in a telephone exchange climbed from a window on to the roof of a double-decker bus.

Cardiff was hit too; Chipperfield's circus had to move lions and other animals.

In Mid-Wales a landslide blocked the A40 road.

The AA reported: 'Innumerable roads are impassable. We are inundated with drivers' calls for help.'

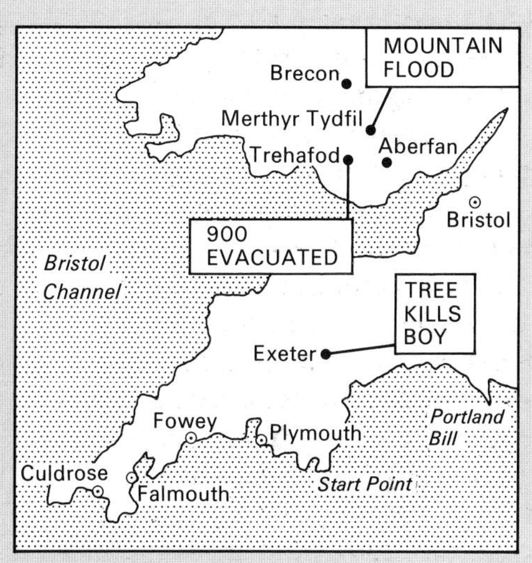

The disaster trail of storm and flood

The Lynmouth floods

On Friday, 15 August 1952 people staying in hotels and guesthouses in Lynmouth were huddled indoors for their tea. They had had a miserable holiday; the weather had been wet for most of the week. As they looked out through the streaming windows at 5 o'clock in the afternoon, they saw that it had started to rain even harder.

Spray from the pouring rain filled the air and soon nothing could be seen through the windows. In ten minutes 40 mm of rain fell and water from blocked gutters poured down the

Fig. 10.2 This rushing torrent was a Lynmouth street

Fig. 10.3 Rivers and rainfall on 15 August 1952 in the area around Lynmouth, Devon

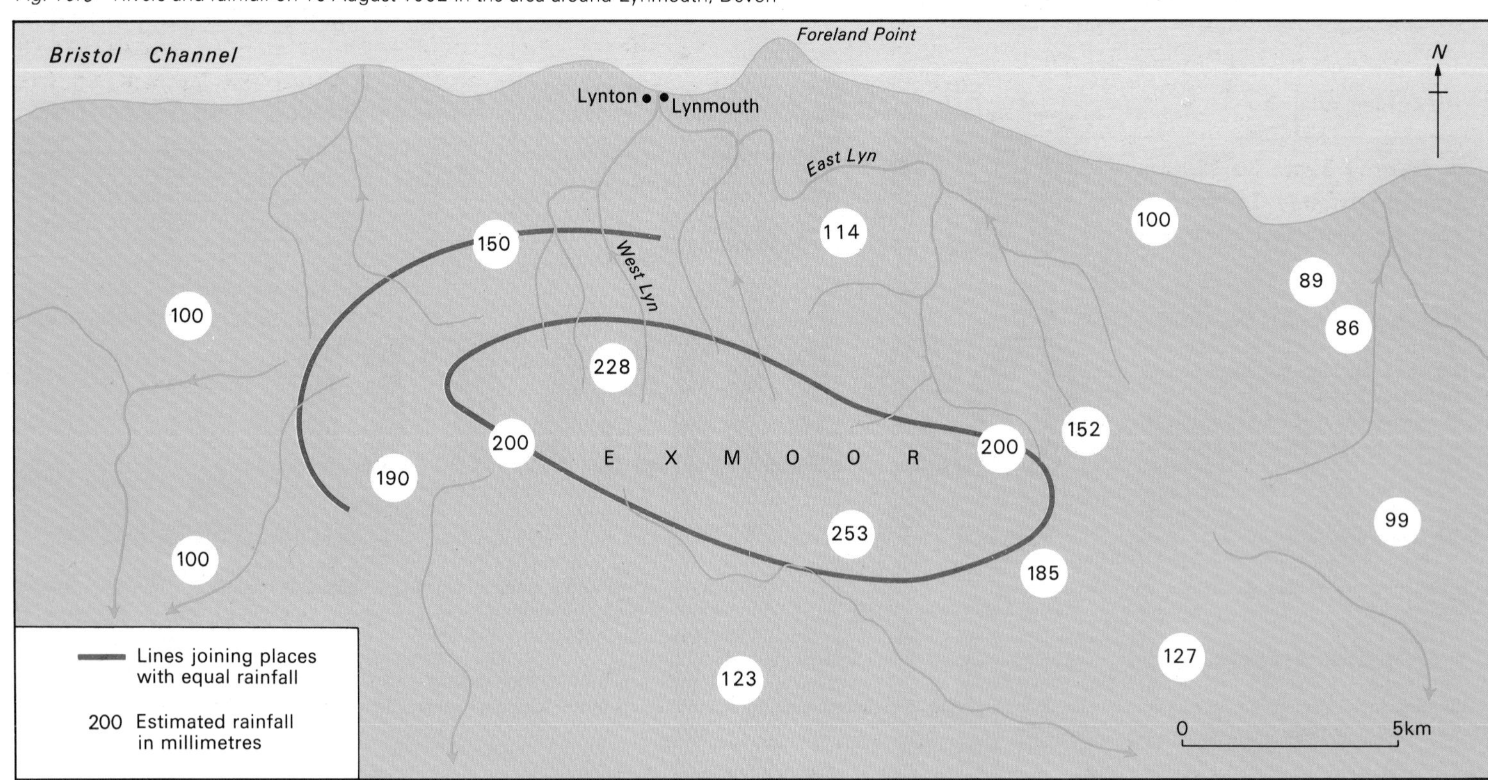

street. Then the rain eased off and the gloomy holidaymakers went back to their tea.

Later that night, as many were going to bed, they heard the rain battering the roof tiles again. Another 40 mm of rain fell in a ten-minute downpour.

2 a) On a copy of Fig. 10.3 draw a line round the catchment area of the East and West Lyn rivers to separate their tributaries from those of the other rivers.
b) The numbers on Fig. 10.3 show the amount of rain that fell on 15 August. There is also a line joining all the places that had 200 mm of rain. On your map shade in the area which had more than 200 mm.
c) A line joining places with the same rainfall is called an **isohyet**. Drawing isohyets is not just a matter of joining up dots. The 150 mm line has been started for you. It has been drawn in by working out where it should be between the dot for 190 mm and that for 100 mm. Draw in the lines in pencil for 150 mm and 100 mm of rainfall.
d) Label your map 'The pattern of rainfall over Exmoor on Friday, 15 August 1952'.

During the night the level of the water in the East and West Lyn rivers rose until the bridge at Lynmouth could not take the flow through its arch. Five hundred cubic metres of water poured down the river every second.

3 a) Use a tape measure to help you work out what a cubic metre looks like. Can you imagine 500 of these flowing past you every second?
b) Work out the volume of your room in cubic metres. How many classrooms full of water flowed down the river at Lynmouth every second?

The amount of water flowing through Lynmouth was roughly fourteen times the average flow of the River Thames at London. Indeed, the Thames has only managed to have a higher flow than 500 cubic metres per second twice in 100 years. In Lynmouth that night 34 people were killed and 90 houses destroyed.

Fig. 10.4 The modern road bridge at Lynmouth has a much wider arch to cope with the river in flood

When it came to clearing up the mess, it was found that 40 000 tonnes of boulders and 250 000 cubic metres of mud had been swept into the little town (Fig. 10.2). Some of the boulders were huge, weighing nearly 10 tonnes. A boulder weighing 7·5 tonnes was found in the basement of a hotel, which tells us that at its peak the water had been flowing at a speed of 9 metres a second.

The river had carried down trees and boulders which formed dams in the narrow sections of the valley. Floodwater had built up behind these dams until they broke, causing a great surge of water to flow down the valley.

When geologists came to look at the valley after the flood, they found that a huge heap of boulders further up the valley had not been moved. They must have been left there by an even greater flood, perhaps the one recorded in 1769. Since then houses in Lynmouth had been built on the floodplain of the river. They clustered round the narrow little bridge with its arch too small to take the water of the river at full spate.

4 Read the account of the Lynmouth flood again and write down as many reasons as you can for the flood being so severe.

5 Are there any parts of your home area that flood at times? See if you can find out when this has happened and perhaps why.

6 Which of the following measures do you think are the most useful for preventing the type of disaster that happened at Lynmouth?
a) Enlarge the bridge arch.
b) Widen the river.
c) Strengthen the river's course.
d) Clear away all fallen trees and other debris from the floodplain.
e) Install a flood warning system upstream.
List the measures in order of usefulness. Give your reasons for this order.

Measuring a river's flow

The flow of water down the River Lyn was not measured by an instrument. It was worked out by measuring the river channel and the level of rubbish left on the valley sides by the flood-waters at their height. The flow of water in a river is measured in different ways depending on how large the river is (Fig. 10.5).

One of the weirs on Fig. 10.5 is shown in Fig. 10.6. It is a piece of plywood which has been placed across a stream so that all the water flowing down the stream must pass through the V-shaped notch in its centre; because of this it is called a **V-notch weir**.

The volume of water passing through the notch in a given time can be measured by catching it in a measuring jug. One can see how long it takes to fill the jug or how much water falls into the jug in half a minute. From this, the **discharge** or flow of the stream in cubic metres per second can be worked out.

On bigger streams a **flume** like that in Fig. 10.7 is used. The instrument in the background measures the depth of water and records it on a drum. In a flume the water is not collected. The discharge is worked out by measuring the cross-section of the water in the flume and multiplying it by the speed at which the water flows through the flume.

Dr Walter Table and his assistant Florence Meter are hydrologists testing the idea that the more it rains, the higher the river rises. They want to know what height the river reaches after 10 mm of rain and perhaps after 20 mm. Then they hope to be able to draw a line on a graph (Fig. 10.8) which will tell them how high the river will flood after 30 mm of rain.

Every time it rained, Florence Meter put on her gumboots and rushed down to the flume to measure the river's discharge. Meanwhile Dr Table put up his umbrella and strolled over to the rain gauge to see how much rain had fallen. Fig. 10.8 shows the amount of rain that fell on each occasion and the increase in depth of the river that resulted.

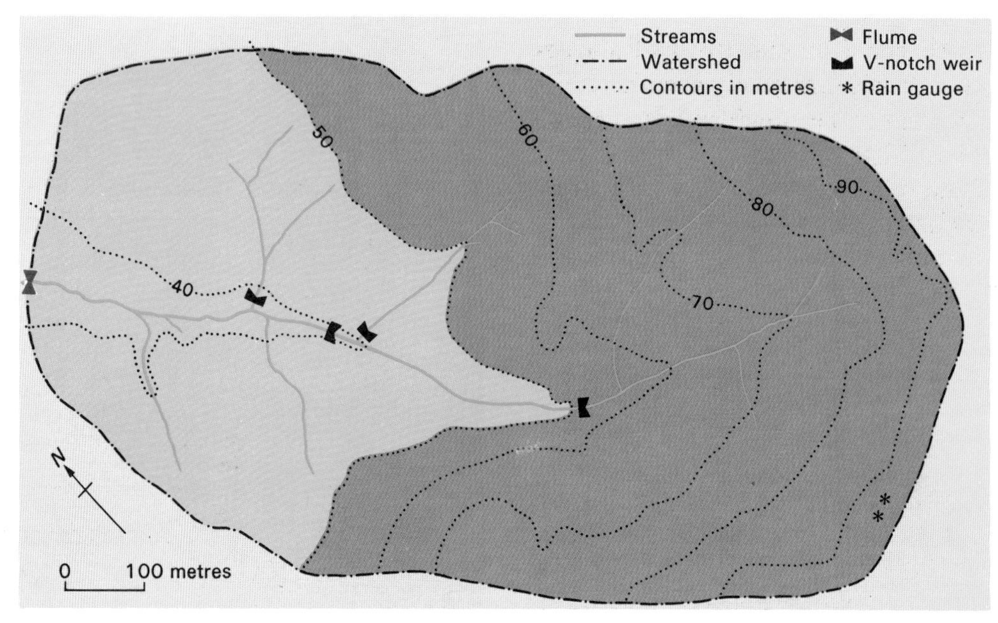

Fig. 10.5 A river drainage basin with instruments for measuring the flow of water

Fig. 10.6 A V-notch weir

Fig. 10.7 A fibre-glass flume

As you can see, the hydrologists were disappointed because there did not seem to be much pattern on the graph. They wondered if the wetness of the soil made any difference to the amount of run-off.

7 a) Put a ruler on Fig. 10.8 so that its edge is over the 10 mm rainfall mark. Use this mark as a pivot.
b) Then slide the ruler gently upwards, keeping the edge on the 10 mm mark until the lowest eight points are covered.

These eight points all had rain falling on dry soil. There had been very little rain on the day before. The water sank into the soil so that there was not much left to run off into the stream. All the other points had rain falling on wet soil so that not much water sank into the soil and more ran off over the surface into the stream.

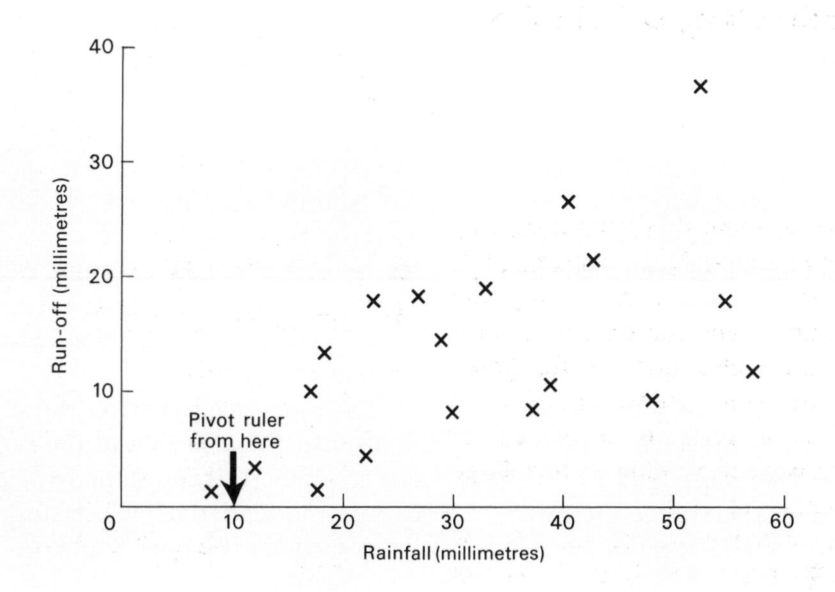

Fig. 10.8 A graph to test the idea that the greater the rainfall, the higher the river rises

8 How can the idea in the last sentence help to explain the severity of the Lynmouth floods described on pages 106 and 107?

Fig. 10.9 Drainage of part of the Lordshill housing estate, Southampton

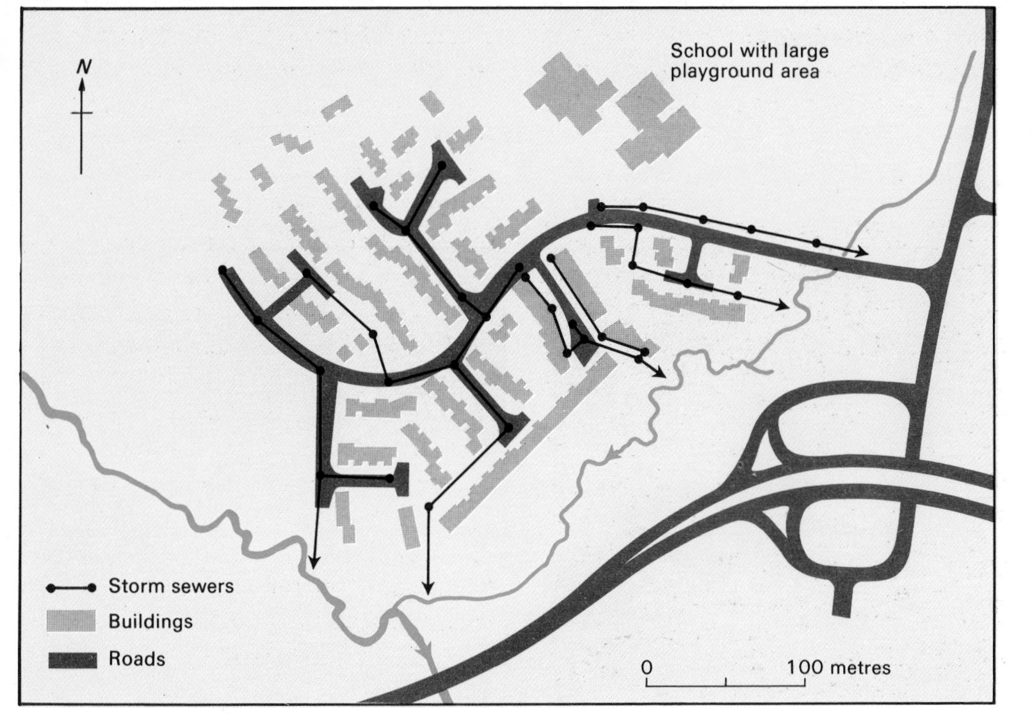

It is not only wet soil that increases the amount of water running off the land. Fig. 10.9 shows a housing estate that was built in Southampton; its roads and buildings do not allow water to sink into the soil.

Instead rain water drains into the rivers along man-made drains. This has led to an increase in the peaks of floods downstream. Stream channels have also grown deeper and wider. The most important effect has been to make floods appear more quickly after heavy rain.

Similar effects were found after some New Towns were built. Large areas of soil were covered in concrete and the number and size of floods downstream suddenly increased.

9 a) On Fig. 10.9 which two types of surface are shown that are likely to increase run-off by not allowing water to seep through?
b) Where would you expect most erosion to take place?

A storm surge

'Down at the sluice, the situation looked dangerous. Barges had been drawn against both sides of the gates and an attempt had been made to buttress the sluice with beams and sandbags, but the piers were bulging dangerously and as fast as material was lowered into the water, it was swept down by the force of the current. The river was foaming over the top of the weir, and from the east, wind and tide were coming up in violent opposition.'

(Dorothy L. Sayers: *The Nine Tailors*, Gollancz)

10 What were the main causes of the flood described in the passage?

In eastern England January 1953 was very wet. Rain and melted snow from the hills inland combined to swell the rivers.

As if this were not bad enough, a depression had made its way across the Atlantic and become stationary over the North Sea (Fig. 10.10). This had two effects. Low pressure causes the sea to bulge up a little at its surface rather like an air cushion does when a fat man gets off it.

The other effect was on the wind. Remember the idea that in the northern hemisphere, if you stand with your back to the wind, the low pressure will be on your left (Chapter 8, p. 90).

☆**11** On a copy of Fig. 10.10 draw in the direction you expect the winds to be blowing.

The gale force winds drove the waters of the North Sea down towards its southern shores, which narrow into a funnel shape. This pushing of the sea into a flood by the wind is called a **storm surge**.

Fig. 10.11 Scenes like this in Norfolk occurred along much of the east coast in the 1953 floods

Fig. 10.10 How the movement of a depression can cause a storm surge

a) The depression has travelled north-east and will pass the north of Scotland

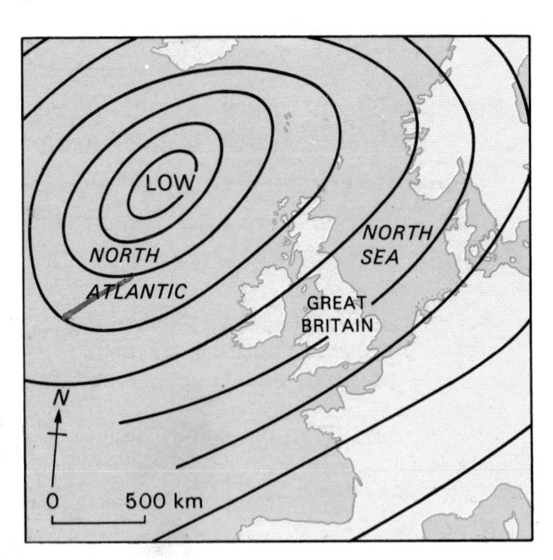

b) The depression is now heading south down the North Sea with the ridge of high pressure heading in from the west.

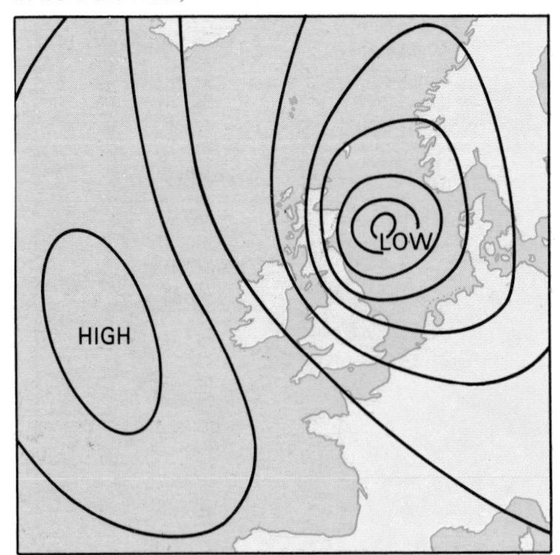

c) The depression has gone over the Low Countries and gale force winds have pushed the surge into the Thames estuary

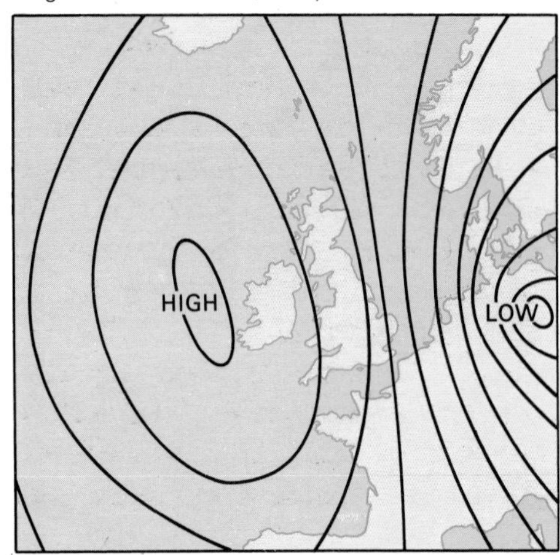

All this combined to produce huge floods on 31 January 1953 which killed 300 people in Britain and 2000 people in the Netherlands (Fig. 10.12).

12 a) Draw a circle in the middle of a piece of paper with six arrows pointing towards it.
b) Label the circle '1953 flood' and the six arrows each with a word or two to sum up the six causes of the flood described above.

Fig. 10.12 Flooded areas along the east coast of Britain, January 1953

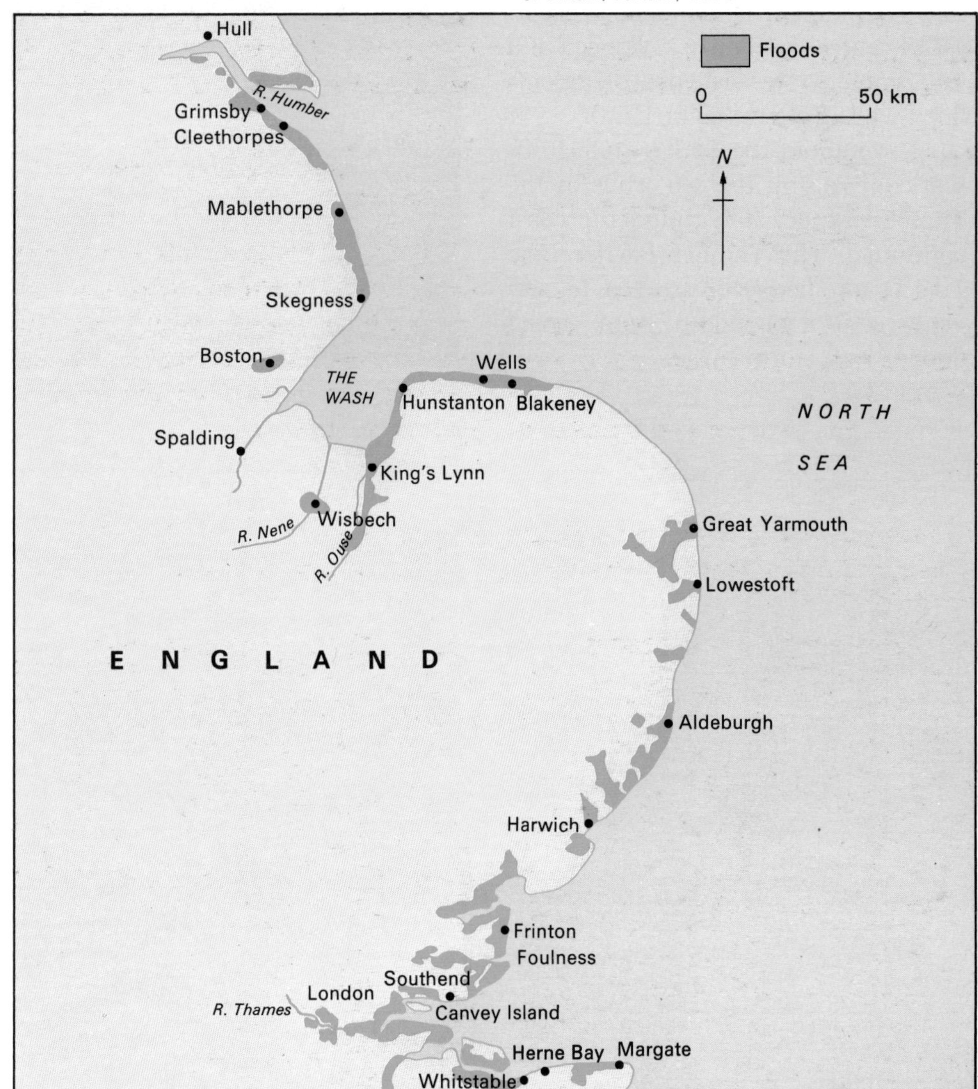

In January the North Sea often has very high spring tides. Also records kept at London Bridge since 1791 show that the height of the tides has been increasing at a rate of about a metre a century.

Because of earth movements, Britain is moving very slowly like a massive seesaw with Hull as the pivot and south-east England sinking. The coastline of the Thames estuary and much of East Anglia is made up of reclaimed marshland. When the silts and peat in this marshland dry out, they shrink and lower the land surface still further. Like the Netherlands much of East Anglia is well below sea-level.

Other floods

This was not the first time floods like the one in eastern England in 1953 had happened. In 1236 the Thames overflowed and men rowed boats inside Westminster Hall. In 1579 the Thames, flooded by melted snow, left fish swimming in the same building.

Pepys's Diary in 1663 tells of the flooding of Whitehall in 'the greatest tide that ever was remembered'. In 1928 central London was flooded and 14 people were drowned. London's flood walls were raised in 1930.

Throughout the 1950s and 1960s work on raising the sea walls in the Thames estuary and round the coast continued. The Thames Barrier (Fig. 10.13) has been constructed to prevent central London from being flooded by a storm surge.

Fig. 10.13 The Thames Barrier has gates which can be raised against a storm surge or lowered to allow shipping through

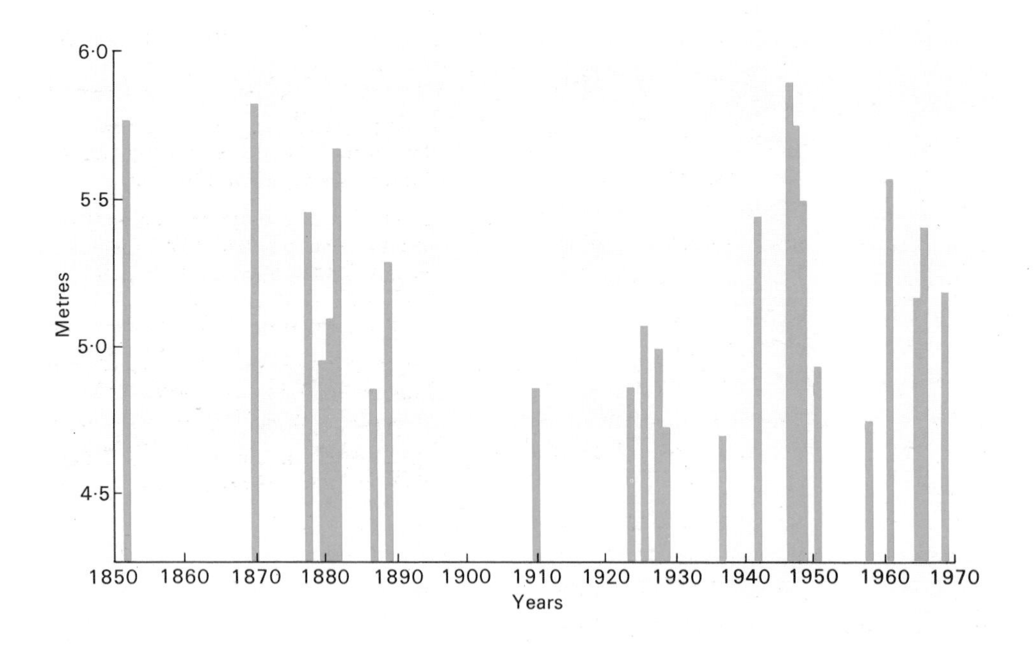

Fig. 10.14 Flood heights of the River Severn at Shrewsbury

Fig. 10.14 shows that the town of Shrewsbury has a long history of flooding.

13 In 1946 the River Severn rose to 5·9 metres (Fig. 10.14). In which year before 1946 did the river nearly reach this height?
b) In 1960 the river reached 5·6 metres. How many times has the river reached that height or above?

14 Old churches and other buildings sometimes have marks carved into them to show the height reached by flood-waters in the past. Do you know of any of these in your area?

Fig. 10.15 The River Ouse overflowing into farmland in 1982

The Vale of York, stretching for some 2500 km² around the towns of York and Selby, also has a history of flooding. It is the floodplain of the River Ouse and of many other rivers that flow into it. Sugar-beet, barley and potatoes are grown on its rich soils.

In Saxon times it was marshland and flooded by the rivers every winter. Gradually over hundreds of years the land has been drained with pumps and ditches. Yet every winter the rivers still tend to flood.

On 6 January 1982 the snow melting on the Pennines and heavy rain made the River Ouse rise 5 metres above its usual level at York. This was the worst flood since 1636 when the river rose 5·3 metres above normal. The Yorkshire Water Authority is now spending £22 million on a 6-km embankment barrier to control the river between York and Selby.

Strangely, part of the trouble seems to be caused by farmers improving the drainage of their land which gets the water out of the fields into the rivers more quickly. When snow melts or heavy rain storms occur, this can result in higher floods happening more often.

The lowland problems are made even worse by changes in the hills around the Vale of York. On the North York Moors, in the Peak District, and on the Pennines the grazing of too many sheep, air pollution, and tourists have all caused erosion of the peat, which holds back the rain water from flowing too rapidly into the rivers. The tourists have trampled the peat and caused disastrous fires. Some landowners have cut ditches to improve the drainage of their moorland rough grazing.

Planting trees might solve some of the problems of these upland areas, but some people think that they should be left undrained as a natural flood storage reservoir and wildfowl reserve.

Nearer the sea there are other problems. The level of the North Sea seems to be rising as the land sinks. Much of the land downstream from Selby is below the level of the highest tides. The peaty lowlands seem to be shrinking as the drained land dries out. The Selby coalfield, opened in 1983, will make large areas of land subside, and it has already had its own flood problems.

The new embankments are planned to be able to cope with floods as high as any since 1831. It remains to be seen if they can manage the floods of the future.

15 Use your atlas to draw a map showing the Vale of York, the North York Moors, the River Ouse, Selby and York.

16 Make a list of as many reasons as you can find why the Vale of York has always flooded at times in the past. Then make another list of reasons for the flooding becoming worse recently.

17 Which of the reasons you have listed for Exercise 16 would you describe as 'natural' and which 'man-made'?

18 What should be done to prevent flooding in the Vale of York?

These are floods caused by earthquake.
Earthquake shock pushes great waves through water.

Tsunamis

One of the types of flood we do not see in Britain is a flood associated with an earthquake. This is called a **tsunami**, which is the Japanese word to describe a tidal wave like the one mentioned in the account of the Lisbon earthquake (Chapter 1, p. 5).

19 Why should the picture in Fig. 10.16 and the name for a tidal wave both be Japanese?

Tsunamis are caused in rather the same way as the wave you produce if you slide down the bath. The shock from the earthquake pushes huge waves through the water. Tsunamis are very dangerous and cause much more damage than the earthquakes themselves.

Thirty-six thousand people were killed by the tsunamis flooding the coasts of Java and Sumatra during the eruption of Krakatoa (Chapter 1, p. 19). However, despite its huge size, no one is believed to have been killed by the eruption of the volcano itself. In the same way, much of the loss of life caused by hurricanes is due to the widespread floods that accompany them.

Fig. 10.16 The Big Wave by Hokusai

Summary exercises

20 Use the last chapter to help you to explain the meanings of the following words:

isohyet	V-notch weir
discharge	run-off
flume	storm surge
hydrologist	tsunami

21 a) Draw a horizontal straight line 20 cm long to make a time-scale. Mark the beginning of the line as AD 1000, the first 2 cm as AD 1100, and so on up to AD 2000. Then mark on the following floods:

AD 1315 Monks in the centre of York row to vespers.
AD1636 River Ouse rises 5·3 metres in York.
AD 1947 Floodwaters around Selby extend for 8 km. They are 2 metres deep.
AD 1982 Ouse at York rises 5 metres.
AD 1983 Storm surge pushes crest of tide to 6 metres above sea-level at Selby.

b) Which of the following facts do you think tend to make flooding worse in the Vale of York? Explain how this might happen.

i) Sudden snow-melt in the Pennines.
ii) Ministry grants available to enable farmers to drain their fields more rapidly into the rivers.
iii) Air pollution, trampling, and overgrazing cause erosion and drying out of upland peat in the Pennines.
iv) Selby coalfield developments could lower the land surface by 1 metre on average in the next 30 years.

11
Pollution

River pollution

In 1845 a man called Friedrich Engels stood on a bridge in Manchester. Here is what he saw:

'At the bottom the Irk flows, or rather stagnates. It is a narrow, coal-black, stinking river full of filth and rubbish which it deposits on the more low-lying right bank.

'In dry weather this bank presents the spectacle of a series of the most revolting blackish-green puddles of slime. From their depths bubbles of foul gases constantly rise and create a stench which is unbearable even to those standing on the bridge forty or fifty feet above the level of the water. Moreover, the flow of the river is constantly interrupted by numerous high weirs, behind which large quantities of slime and refuse collect and putrify.

'Above Ducie Bridge there are some tall tannery buildings and further up there are dye-works, bone mills and gasworks. All the filth, both liquid and solid, discharged by these works finds its way into the River Irk which also receives the contents of the adjacent sewers and privies.'

1 a) Which signs of pollution mentioned in the extract can you see in Fig. 11.1?
b) List any other types of pollution shown in Fig. 11.1 but not mentioned by Engels.

The pipe pouring sewage into the River Irwell in Fig. 11.1 will cause

Fig. 11.1 The view from Blackfriars bridge, Manchester in 1876

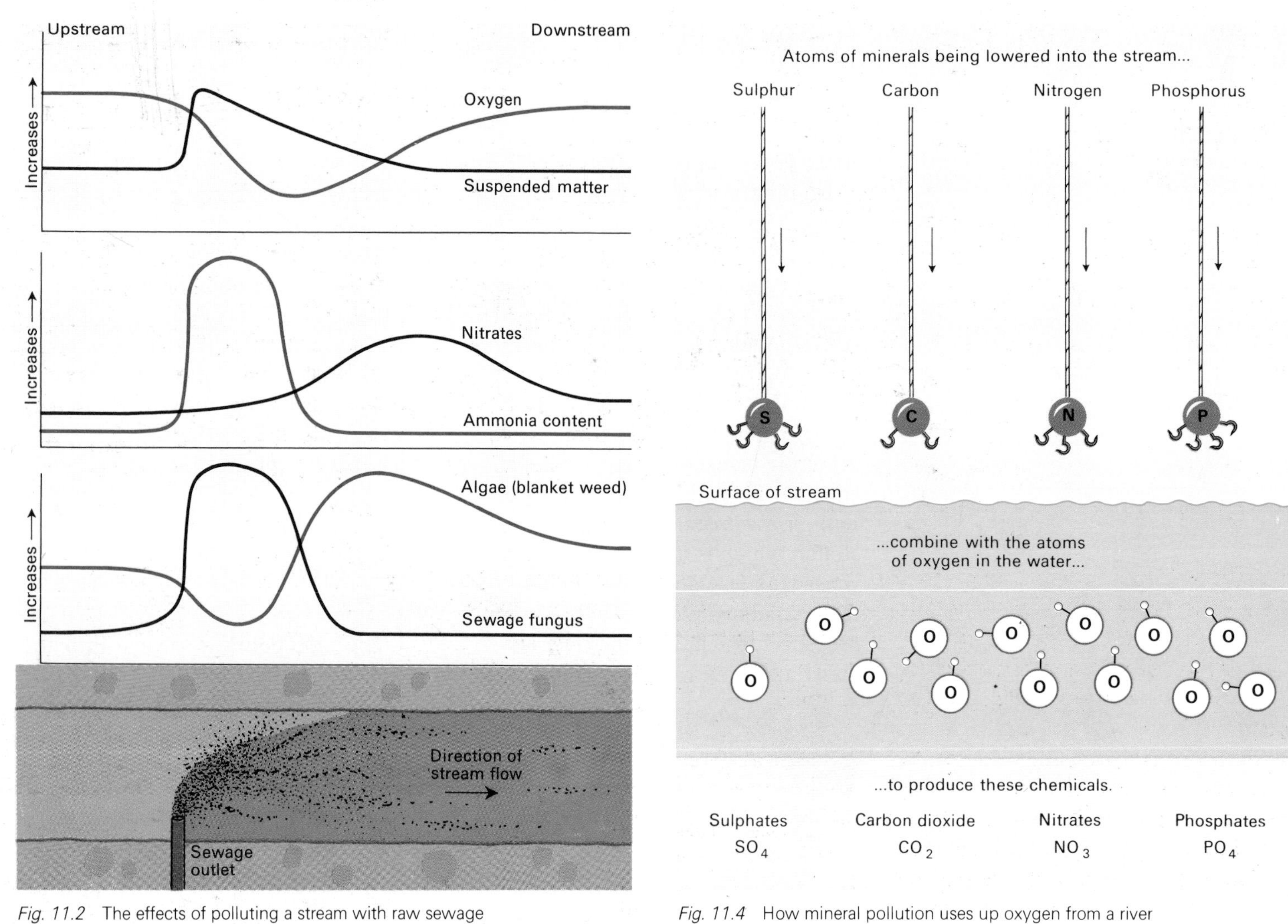

Fig. 11.2 The effects of polluting a stream with raw sewage

Fig. 11.4 How mineral pollution uses up oxygen from a river

Fig. 11.3 Water creatures that show how polluted the water is

Sludge worm

Caddisfly larva

Rat-tailed maggot

Blood worm

Water louse

Stonefly nymph

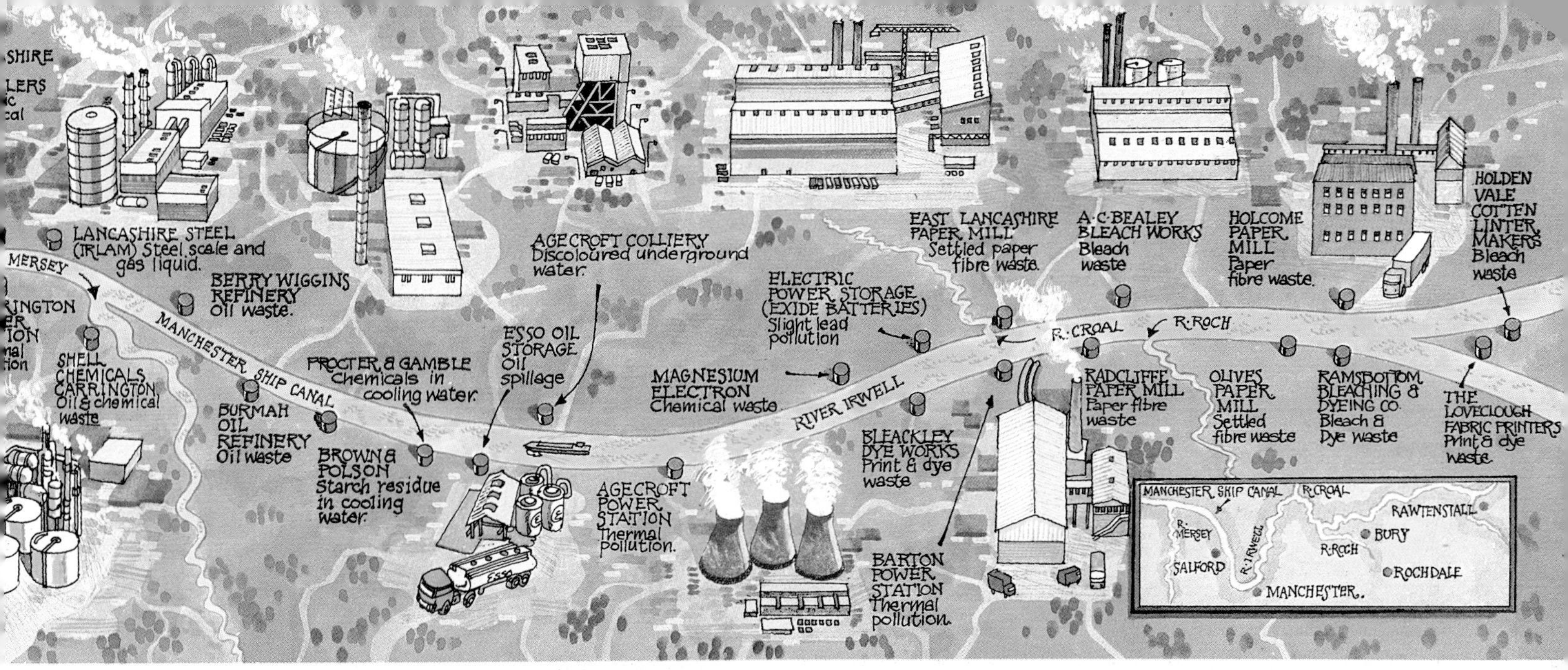

Fig. 11.5 Pollution from factories along the River Irwell in 1970

changes in the stream. Fig. 11.2 shows the effect of such pollution. Upstream the water is clear and fresh; there is lots of oxygen in the water for all the water creatures in Fig. 11.3 to breathe.

As soon as the sewage enters the water, bacteria get to work, breaking the sewage down into ammonia. To do this they use up oxygen which they take from the water. If all the oxygen is used, only animals with breathing tubes like the mosquito larva and the rat-tailed maggot can survive, together with sludge worms and tubifex worms which live on dead plants and animals (Fig. 11.3).

Other bacteria break down the ammonia into nitrates which are a rich plant food. Green pond weed or algae will grow rapidly until the water turns bright green like pea soup. Decaying plants use up still more oxygen. In this part of the river, besides the rat-tailed maggot and sludge worm, there are bloodworms and water lice.

Further downstream more sensitive animals begin to appear again: caddisfly larvae in their little cases of twigs and sand, and the stonefly nymph.

2 Make a copy of Fig. 11.2 and divide it up into zones:
a) very polluted.
b) quite polluted.
c) the normal stream.

When minerals are tipped into a river, oxygen is again used up to deal with the pollution (Fig. 11.4).

☆3 a) Use wooden or polystyrene balls to make models of the atoms of minerals and oxygen in Fig. 11.4, fitting hooks and eyes into them as shown.
b) Attach a piece of string to each mineral atom. Hang these over the side of a table representing the surface of the stream.
c) Under the table in the 'stream', attach oxygen atoms to the mineral atoms, one on each available hook.
d) Pull up the strings and see what sort of mixtures of atoms result. List them.
e) What has happened to the oxygen atoms in the 'stream'?

Fig. 11.6 Pollution from factories along the Irwell

If only one pipe pollutes a river, the water can clear itself within a kilometre or so. As we can see from Engels's description and from Fig. 11.1, if too many pipes pollute a river, the river dies. Although Engels lived in 1845, not much changed until very recently (Fig. 11.5).

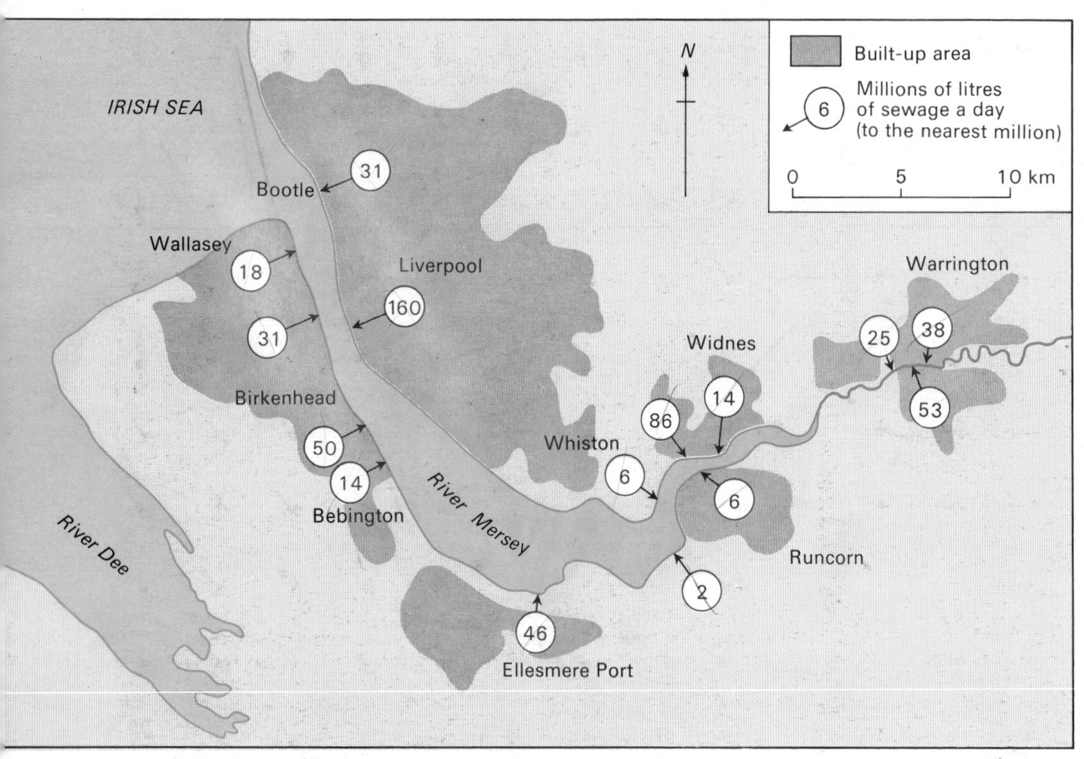

Fig. 11.7 Pollution on the Mersey: sewage and industrial waste

Fig. 11.8 River Mersey: biochemical oxygen demand (BOD)

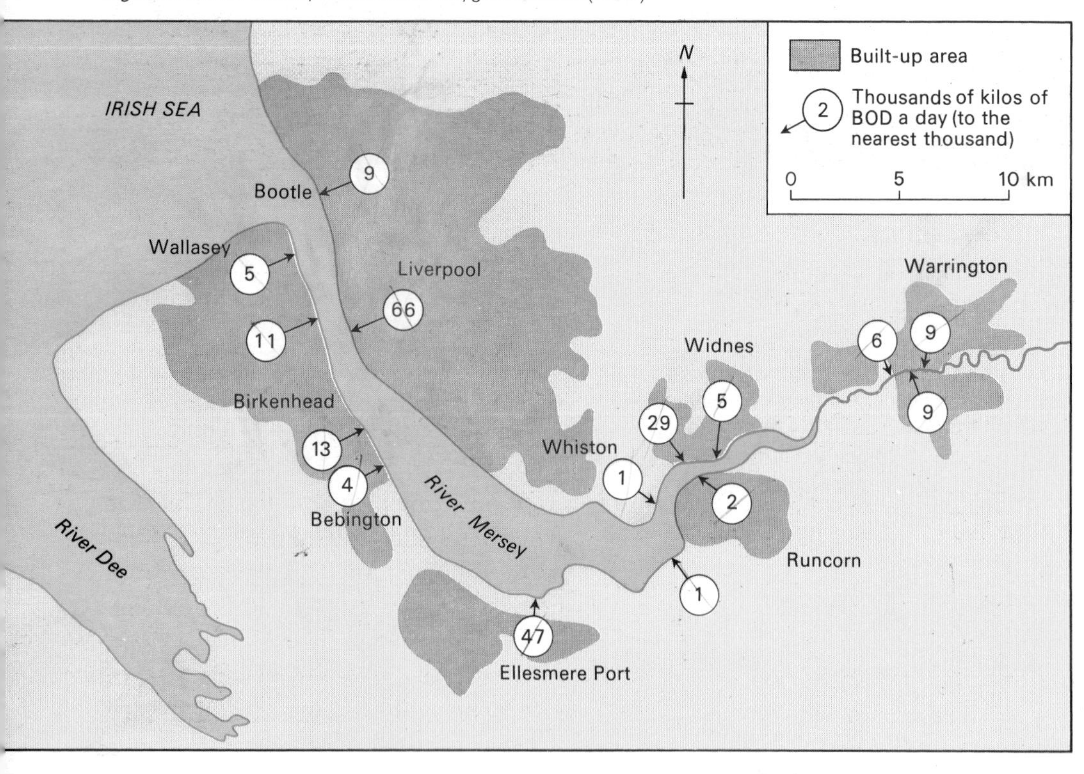

Fig. 11.7 shows the pollution of the River Mersey in 1972. Bacteria trying to cope with the pollution use up the oxygen. One way of showing the oxygen shortage which results in the water is to work out the biochemical oxygen demand (BOD) (Fig. 11.8).

☆**4** a) Use the figures in Figs. 11.7 and 11.8 to draw a scattergraph. Show the amount of sewage and other industrial waste pumped into the river on the horizontal axis and the BOD on the vertical axis.
b) Is it true to say that the greater the amount of pollution, the higher the BOD?

The River Thames (Fig. 11.9) used to suffer like the Mersey. Yet now the waters of the Thames can support salmon and other freshwater fish. Councils have spent money improving sewage works and firms have spent more money on treating their wastes instead of tipping them into the river. The Thames is no longer a dangerous running sore and has become a pleasure to live by.

The North West Water Authority is building modern sewage works to treat the sewage before it reaches the River Mersey. The Control of Pollution Act of 1974 has attempted to cut down the amount of tipping by factories and the Merseyside Council is planning to spend the next twenty years cleaning up the river.

Fig. 11.9 A nineteenth-century drawing of Death being carried by the polluted waters of the Thames

Fig. 11.10 Fishing in the River Thames today

5 Are the streams in your area polluted? A simple survey at points down a river with a pond-dipping net, a white dish, and a bucket will reveal what sort of water creatures live in the stream. (Note: take great care not to fall in the river; banks can be slippery and crumble easily. Polluted rivers are dangerous and may be deeper than you think.)
a) Fill the dish with river water. Sweep the net along the bottom of the river and empty the contents into the dish.
b) How many different kinds of water creature have you found? Fill in a table like Fig. 11.11. Then empty the contents of your dish into the bucket.
c) Do this three times to help you decide how polluted the section of river is.
d) Put the animals back in the stream when you have finished. You do not want to leave the stream completely without water life.

Water creatures	First net sweep	Second net sweep	Third net sweep
Mosquito larva			
Rat-tailed maggot			
Sludge worm			
Tubifex worm			
Bloodworm			
Water louse			
Caddisfly larva			
Stonefly nymph			

Fig. 11.11 Table for water pollution survey

Polluting the sea

In March 1967 the tanker *Torrey Canyon*, carrying 118 000 tonnes of crude oil from Kuwait, ran aground on the Seven Stones Reef off Land's End in Cornwall. For ten days, while the government waited to see if the oil could be pumped off, oil leaked into the sea (Fig. 11.15).

Thirty thousand tonnes drifted up the English Channel, 20 000 tonnes reached west Cornwall, and another 50 000 tonnes floated into the Bay of Biscay. By the time the government took action only about 18 000 tonnes of oil were left on board.

Attempts to burn it with bombs and napalm failed. Vast amounts of detergent were sprayed on the sea to disperse the oil, but the effect of the detergent itself on wildlife was disastrous. Twenty thousand guillemots and 5000 razor-bills were killed. Of the 8000 birds rescued, only 450 were still alive by the middle of April.

The oil floating towards the French coast was not treated with detergent. It was left to be dispersed by the wind, waves, and tides. Because of the big distance from Land's End, no French beaches were polluted.

About ten years later another tanker, the *Amoco Cadiz*, ran aground on the rocks of Brittany (Fig. 11.12). She was carrying 220 000 tonnes of crude oil. We can predict where the oil will reach if we know the pattern of winds. Follow the directions below and then do the exercises to see how close is your pattern of oil pollution to the one shown in Fig. 11.12.

Fig. 11.12 Based on a report in the *Financial Times*, 23/3/78

Worst ever oil pollution

By Mark Webster **Brest, March 23**

The grounding of the Amoco Cadiz is now the biggest pollution disaster in tanker history. Oil is still leaking fast from the stricken ship and there is little hope of pumping out any of the remaining oil.

Battered by gale-force winds, the last of the giant tanker's 14 storage tanks has ruptured and is leaking oil.

The pollution disaster has affected more than 60 miles of French coastline which includes fishing ports and holiday areas. Last night a 'breakaway' oil slick was reported about 50 miles from the Channel Islands but a forecast of northerly winds suggests that France will have the Amoco's entire cargo driven on to its shores.

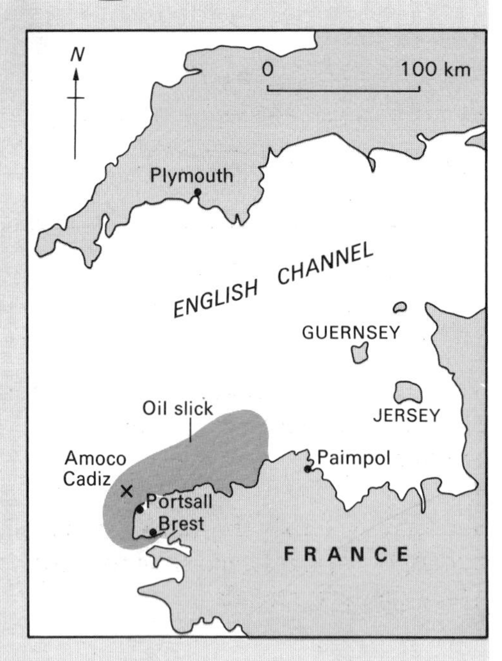

1 Trace the arrows showing the direction of the wind from Fig. 11.14. Put your tracing over a copy of Fig. 11.13 to show the position of the centre of low air pressure at the time the tanker ran aground. This will show the direction of the winds at the time the oil began to leak.

2 Shade the square with the wreck to show the first leak of oil. Now throw a dice. Use the little table on Fig. 11.13 to show you how to move the tracing to see where the low pressure has moved to six hours after the mishap.

3 You can tell which way the wind is blowing the oil slick by looking at the direction of the arrow now over the wreck. If it is pointing towards the coast, the oil will wash ashore; if it is pointing any other way the oil slick will be blown one square in that direction.

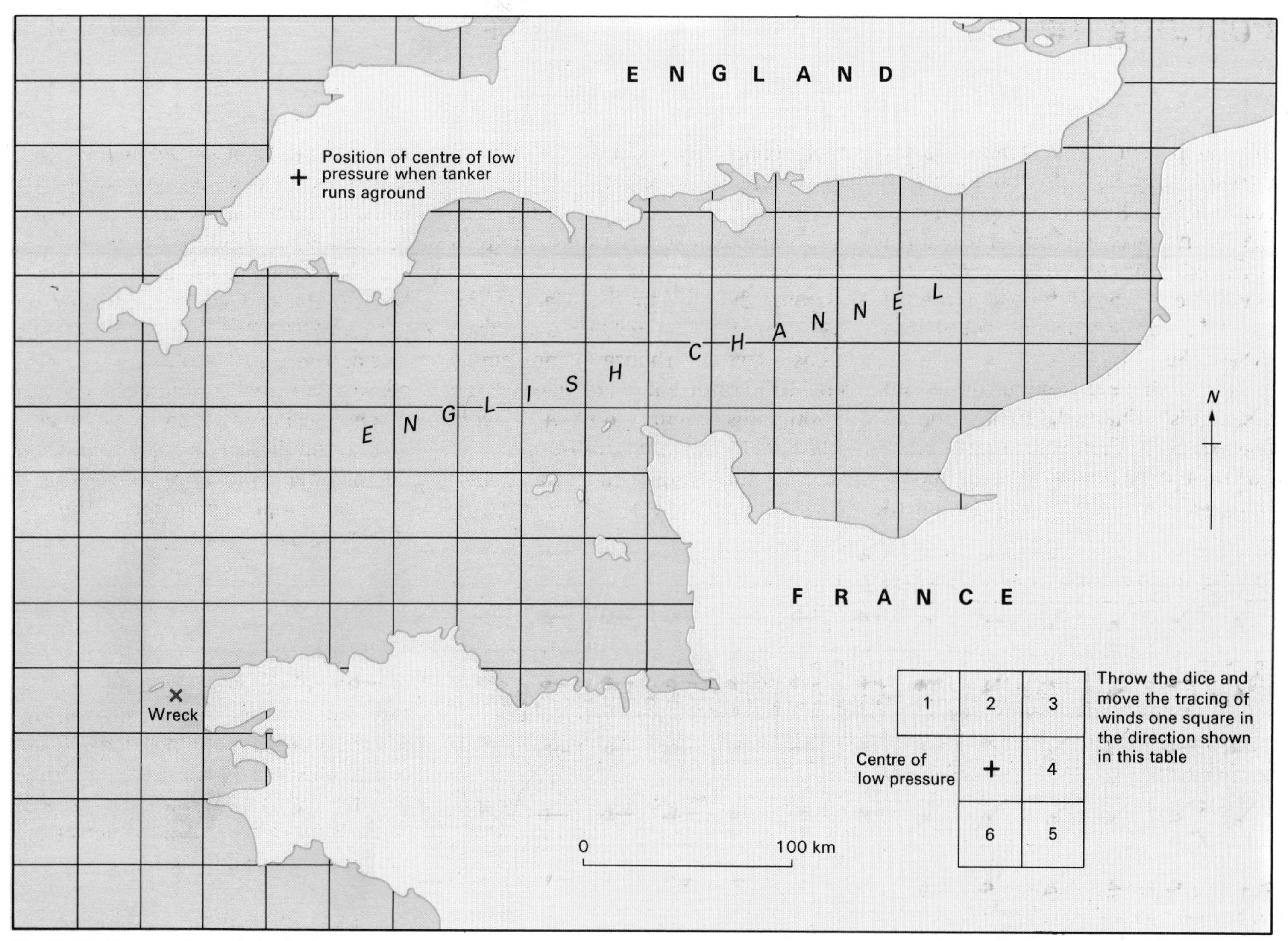

Fig. 11.13 Base map for coastal pollution game

4 Shade the square on your map to show how far the oil slick has moved in six hours. Now both your shaded squares will contain oil to be blown by the wind.

5 Throw the dice again and move the centre of low pressure according to the table. Look at the two shaded squares and at the arrows covering both of them. Shade in the two squares to which the arrows are pointing.

6 Carry on for six throws of the dice, shading in squares each time to show where the slick spreads.

6 a) Compare your map with the one in the newspaper extract (Fig. 11.12); is your oil slick spreading in the same direction?
b) Carry on for another six throws and see how far the oil has reached before it begins to settle 72 hours after the wreck. The oil will spread no further but will become 'tar' on the beach.

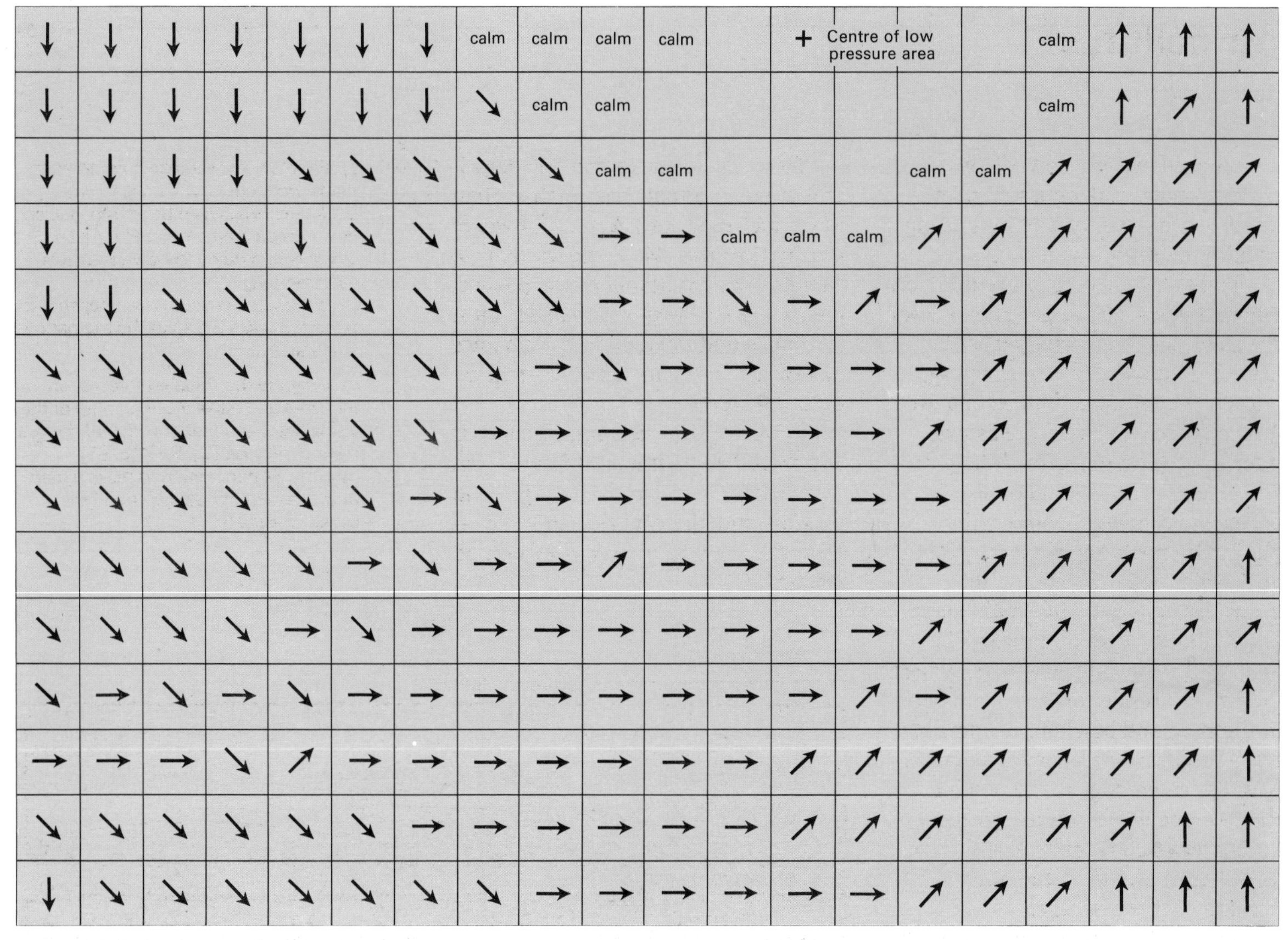

Fig. 11.14 Pattern of winds at the time the tanker ran aground

☆**7** Choose another place for the wreck, say on the English coast, and see where the oil would have reached in 72 hours.

☆**8** Turn your tracing paper over and change the centre of low pressure to a centre of high pressure. Now see what happens when, instead of a depression, high pressure is centred over England.

Fig. 11.15 Oil slick spreading from the wreck of the *Torrey Canyon*

Air pollution

'... Smoke lowering down from chimney-pots, making a soft black drizzle, with flakes of soot in it as big as full-grown snowflakes ...

'Fog everywhere. Fog up the river, where it flows among green meadows; fog down the river, where it rolls defiled among the tiers of shipping, and the waterside pollutions of a great (and dirty) city. Fog on the Essex marshes, fog on the Kentish heights. Fog creeping into the cabooses of collier-brigs; fog lying out on the yards, and hovering in the rigging of great ships; fog drooping on the gunwales of barges and small boats.

'Fog in the eyes and throats of ancient Greenwich pensioners, wheezing by the firesides of their wards; fog in the stem and bowl of the afternoon pipe of the wrathful skipper, down in his close cabin; fog cruelly pinching the toes and fingers of his shivering little 'prentice boy on deck.

'Chance people on the bridges peeping over the parapets into a nether sky of fog, with fog all around them, as if they were up in a balloon, and hanging in the misty clouds. ...

'... Most of the shops lighted two hours before their time—as the gas seems to know, for it has a haggard and unwilling look.'

(Charles Dickens: *Bleak House*)

9 a) What are the main effects of fog described in the extract?
b) What signs can you find that the fog carries smoke and soot?
c) What hints can you find that Dickens thought that fog was unhealthy?

The official definition of **fog** is when you cannot see further than 1000 metres (1 km). You and I would not notice this very much, but to captains of ocean tankers and pilots of planes fog is a very real danger; most jet planes and tankers need a distance of at least 3 km in which to pull up from full speed.

Thick fog is when the visibility is less than 200 metres; this happens on more than ten days in a year in London and the Midlands and on at least five days in a year elsewhere in Britain.

10 a) From a good vantage point in your home work out how far it is to any landmarks which can be seen on a clear day. Make a diagram like Fig. 11.16. Use this as a guide to how foggy it is each morning.
b) If everyone in your class does this, it is easy to make a map of the fog pattern in the area.

11 a) Listen to the shipping forecast on the radio and make a careful note of the visibilities mentioned for each of the sea areas shown on Fig. 8.13 (p. 91).
b) Use this information to make a map of the pattern of visibility round the coasts of Britain.

Fig. 11.16 Landmarks useful for working out visibility

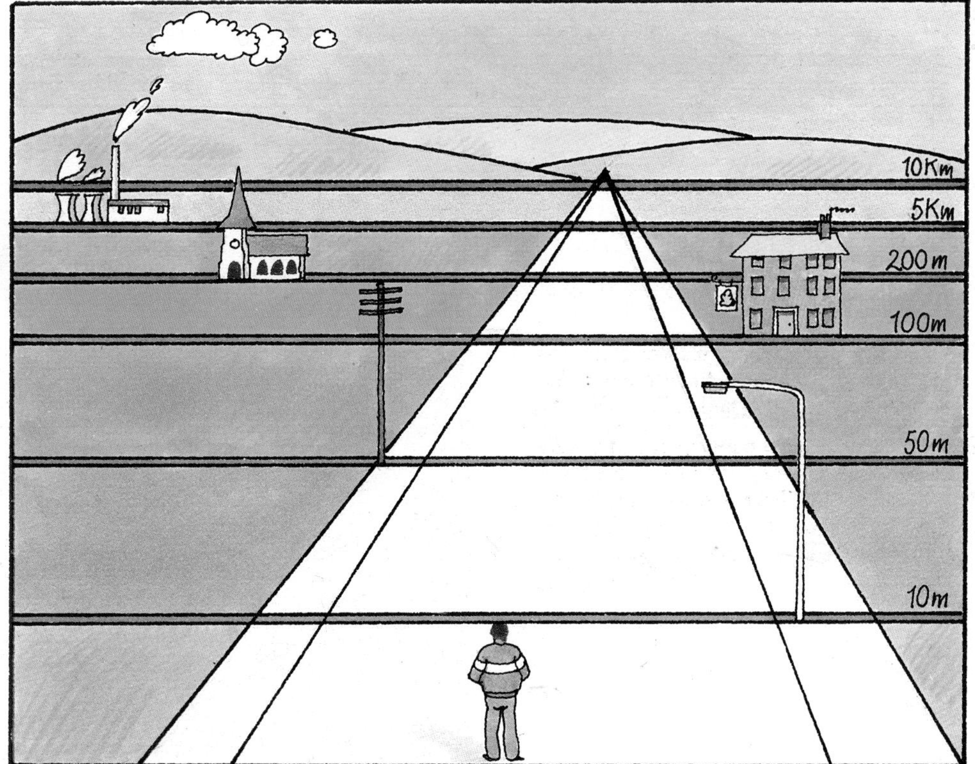

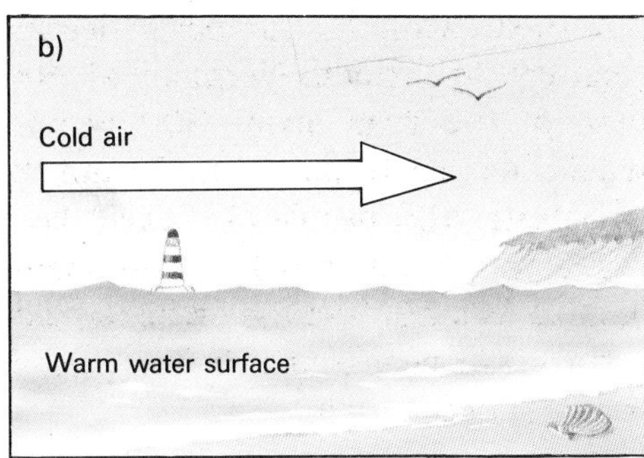

Fig. 11.17 How fog can form

There are four types of fog:

i) Under clear night skies the ground cools to the point where the water vapour in the air just above condenses; this is called **radiation fog.** Most fog in London and the south of England is caused in this way.

ii) When moist air blows over cold or snow-covered ground, the result is **advection fog**. This is more common in the west of Britain.

iii) **Steam fog** is caused by cold air passing over warm water such as might be found downstream from a power station or a factory. It is sometimes found in north-west Scotland where cold air drifts over warm water currents in the sea.

iv) **Hill fog** is simply low cloud.

12 Fig. 11.17 shows the main causes of fog. Use the four descriptions above to work out which diagram shows which kind of fog. Copy the diagrams and add captions.

13 On days when it is foggy in your area, try to decide which kind of fog it is with the help of Fig. 11.17. This is often best done in November.

Fog is natural and consists of white, pure water droplets in the air. The kind of fog described by Dickens was not natural or healthy. During the nineteenth century so much coal was burnt that the smoke and soot turned houses and buildings black. It was only in 1952 that some people began to realize just how harmful fog is to health.

In early December of that year air pressure was high and there were few winds blowing. The Thames valley was filled with cold still air to a depth of 100–200 metres (Fig. 11.19). Above this the air was warmed by the sun, but the smoke and fog in London stopped the sun's rays reaching the ground.

Fig. 11.18 A 1940's newspaper cartoon by Giles

"War, fog, international chaos—nearly time you men did something about it."

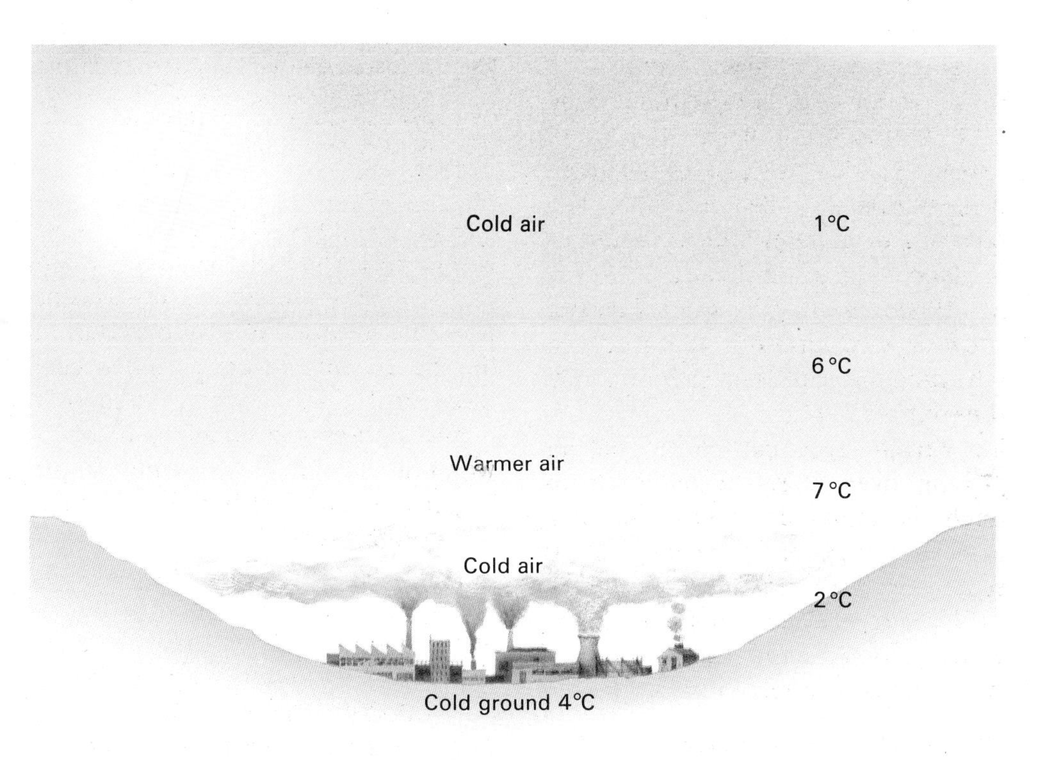

Fig. 11.19 Cold air trapped in the Thames valley causing smog

Fig. 11.20 Deaths in London caused by smog

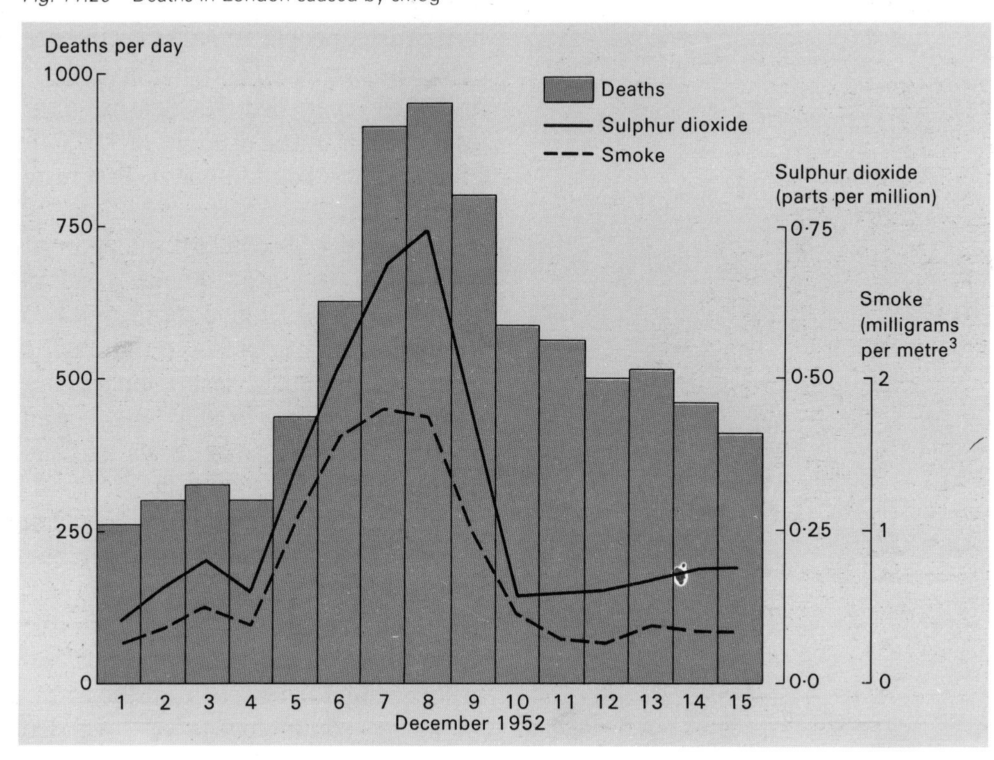

The air stayed cold and filthy with no wind to blow it away. On 6 December visibility at London Airport was less than 12 metres. The soot and sulphur dioxide from London's chimneys filled the fog and turned it into deadly **smog**.

How deadly the smog was can be seen from the bars on the graph (Fig. 11.20) showing the number of deaths each day in London. Normally at this time of the year about 300 people would die each day. On 7 and 8 December, 900 people died each day, particularly from bronchitis and pneumonia. As a result of the awful fog, an extra 4000 people died in the period shown on the graph.

14 Try an experiment to see how much dirt is carried in the rain.
a) Put a filter paper in a funnel supported by a rain gauge or even a milk bottle.
b) After a rainy day look at the filter paper under a microscope and see how many specks of soot or ash you can see.

15 Another way of seeing how much dirt is dropped from the air is to leave a sheet of white card or paper in a corner of the room. Place twenty coins on the paper and remove one each day. Do this in a room with the door closed for most of the time and the window slightly open.

The Clean Air Act of 1956 forbids the burning of coal other than smokeless fuel on domestic fires and the emission of black smoke from factory chimneys in areas which have been made into clean air zones by the local council.

Sulphur dioxide in smoke from chimneys will turn to dilute sulphuric acid in rain, and this can have ill-effects on soils, plants and water supply. Chimneys giving off sulphur dioxide either have to filter it out or

Fig. 11.21 Plumes of pollution blowing north-west from Katowice in Poland

Fig. 11.22 Fog can be a great hazard for the motorist

they have to be very tall so that the pollution is spread thinly over a wide area by the time it reaches the ground.

Yet, viewed from satellites, plumes of smoke can still be seen drifting across Europe (Fig. 11.21). While Britain may not suffer from this pollution, it may harm the rest of Europe. Fig. 11.23 shows how the rainfall in Europe has become more acid since 1956, when measurements began.

16 Look at Fig. 11.23 carefully. Where does it seem to suggest that much of the pollution comes from?

While most of our cities have cleaned the soot from their air, a new danger seems to be appearing. Sunlight acts on exhaust fumes from motor vehicles to produce a foul yellow haze called **photochemical fog**.

In places as far apart as San Francisco, New York, and Los Angeles in the USA, and Milan in Italy this fog has caused people to suffer from sore eyes, runny noses, coughs, bronchitis and has even led to deaths from pneumonia. In the drought of 1976 in June and July London had its first taste of this hazard.

Other worries concern the amount of lead in the atmosphere. Lead is added to petrol to give more miles to the gallon and to preserve the engine from wearing out so fast. People living near large busy roads may well breathe in dangerous levels of lead pollution from car exhausts.

Lead pollution could also be taken in by eating food such as vegetables and fruit grown close to a main road. The lead may be deposited from the atmosphere on the leaves of plants and absorbed by their roots from the soil.

Many climatologists are worried

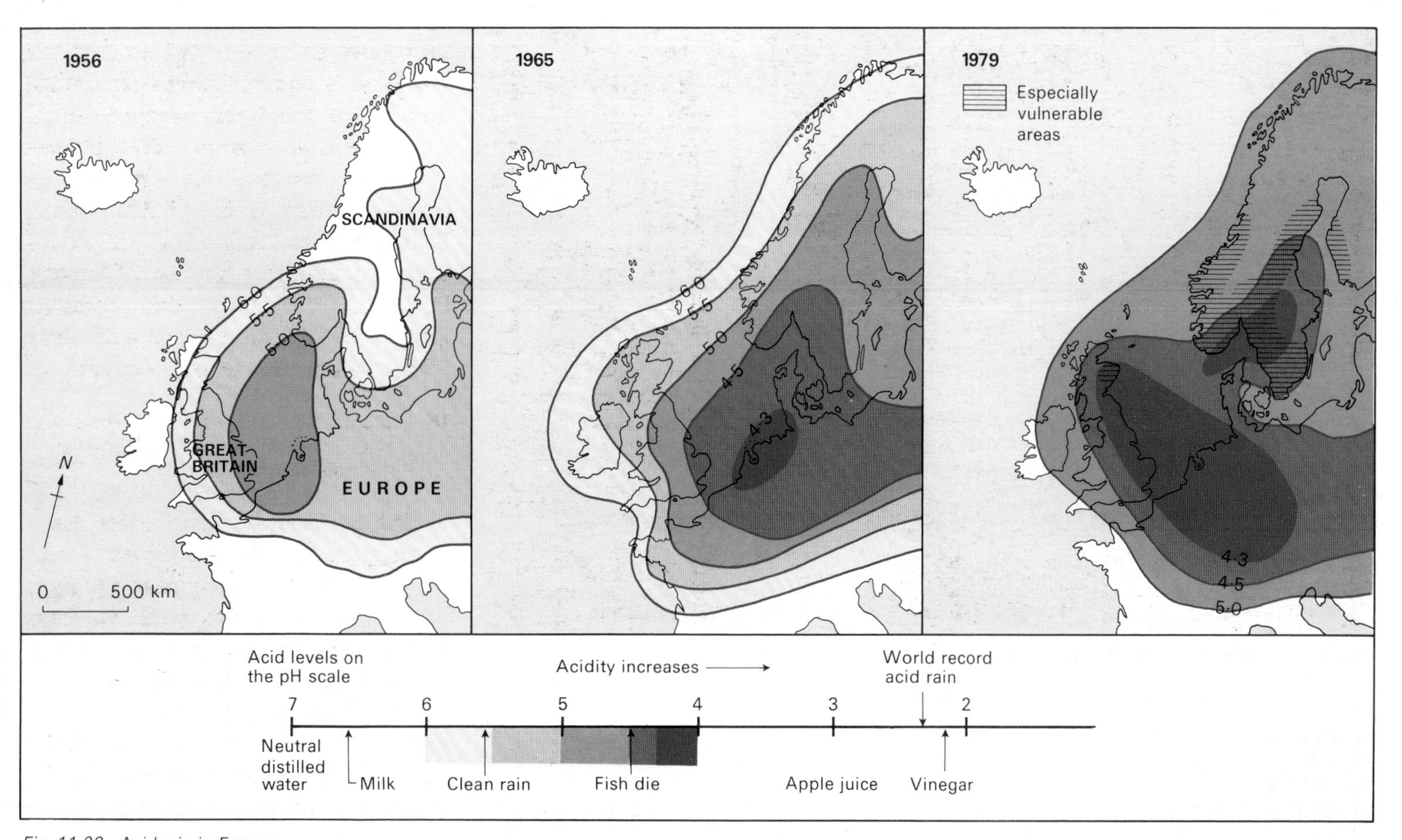

Fig. 11.23 Acid rain in Europe

about other forms of air pollution. Some, for example, think that chemicals used in aerosol sprays may be drifting up into the upper atmosphere, where they damage the ozone layer and cause an increase in world temperatures. Others suspect that an increase in the amount of dust in the air from soil erosion, sooty chimneys and volcanoes may be having the opposite effect and cooling the earth's climate.

Just as with water pollution, we may not have progressed as far since the nineteenth century as we think we have.

17 Make a table with two columns, the first listing the types of air pollution discussed in this section, and the second describing the effects of each type.

Summary exercises

18 Use the last chapter to help you to explain the meanings of the following words:

fog	hill fog
thick fog	smog
radiation fog	photochemical fog
advection fog	acid rain

19 On your copy of Fig. 11.2 mark on where you would expect to find the various water creatures in Fig. 11.3.

20 Use Fig. 8.8 (p. 88) to predict which shores would be threatened by massive leaks from oil wells:
a) in the North Sea.
b) in the Persian Gulf.

Index